PERFECT TECHNOLOGY SECRETS

Unlock the Business of Your Dreams with
Perfect Technology

DR. ADAM LINK

Perfect Technology Secrets
© 2022 Adam Link

First published in 2022 in paperback
in the United States of America
by More Light Press / More Light LLC.

More Light Press
From the Iron Range in MN

Printed in the United States of America

ISBNs
Print Edition: 978-1-959850-00-7
Electronic Edition: 978-1-959850-01-4
Hardcover Edition: 978-1-959850-02-1

First Edition

Table of Contents

Introduction.. v

Section One 1

Chapter 1: What Is The CHESS™ Framework?................. 3

Chapter 2: Discovering The CHESS™ Framework 17

Chapter 3: What Can CHESS™ Do For My Business? 29

Section Two...29

Overview.. 30

Chapter 4: Technology Does Not Fit My Business Goals 43

Chapter 5: Five Families of Technology™ 49

Chapter 6: Business Goals and Differentiation 79

Chapter 7: Business Archetypes Explained...................... 89

Chapter 8: Wrapping Up "Technology Does Not Fit My Business Goals" 105

Chapter 9: Technology Does Not Fit My Business Culture 115

Chapter 10: Business Processes And Technology........... 119

Chapter 11: The Two Critical Technology Frameworks 127

Chapter 12: Wrapping Up "Technology Does Not Fit My Business Culture" 141

Chapter 13: I Do Not Know What I Do Not Know......155

Chapter 14: Plan Do Check Act – The Evaluation Cycle 175

Chapter 15: Wrapping Up "I Do Not Know What I Do Not Know" 183

Chapter 16: The Technology Factor 195

Section Three 211

Chapter 17: The CHESS™ Assessment Online...............213

Chapter 18: Applying CHESS™ To Your Business' Technology Choices..221

Chapter 19: Relationships Between The Four Variables.231

Conclusion 238

INTRODUCTION

I f you picked up this book, you have a nagging feeling that the technology in your business could be better. You may not know exactly what is wrong, but you have this feeling that there is a better way to run your business using technology. If this is you, you are one of us. The pursuit of better technology to run one's business more efficiently, profitably, or to grow the size of your empire is a journey many of us are on.

You are a Tech Seeker. A Tech Seeker is someone who believes there is a better way to use technology in their business. My own journey began over 10 years ago when I realized that my business was not operating as efficiently as it could have and that I was spending too much time focusing on the technology and not the business.

If you have this feeling that technology could be better for your business, then you have found the right book. In these pages, I will reveal the secrets that I have learned of how to align your business and technology to achieve your business goals. I have used these concepts successfully to grow and sell three different companies. I have used these concepts as a part of the team that took a company public through an $80 billion IPO. I learned these concepts through failing in business multiple times as well. The secrets contained in this book are the result of over a decade of real-world experience - successes and failures. These secrets are also the direct result of my own academic research spanning nearly a decade as I uncovered insight into the way that business owners, like yourself, use technology in their business to achieve their business goals.

By the end of this book, you will have a framework that you can use to achieve your business goals faster with technology. While this book is primarily written for the small business owner, the concepts are equally applicable to a large enterprise or even a nonprofit. If you have a goal for your business and you want to use technology to achieve that goal faster, then this book will give you the answers you seek.

This book should be thought of as a reference book. While it is perfectly acceptable to read it once and put it aside, it is much more beneficial to read this book with a pen or highlighter in hand, marking the pages with your own insight as to how the concepts can be applied to your business. My hope is that you keep this book close by on your desk as you run your business.

The concepts in this book are evergreen. This means that I do not mention specific technologies, but instead, provide a framework that is applicable to your business in general. Through a decade of experience and academic research and built on knowledge established as far back as 100 years ago, the concepts in this book will set your business up for success using technology of any type into the next century.

So come join me on this journey and learn how technology can help you achieve your business goals faster.

SECTION ONE

Chapter 1

WHAT IS THE CHESS™ FRAMEWORK?

When I first began the journey of creating the CHESS™ Framework, I struggled with the fact that there seemed to be so many different components of technology that needed to be evaluated to understand if that technology was going to work for my business. The more I grappled with this problem, the more I realized that there was actually a logical thread that every business owner took when evaluating if technology was going to work or not. But there were also some specific characteristics of each business owner that I talked to that helped me understand their general attitude towards technology. Further, the attitude towards the technology depended on the technology itself.

Combining my academic research with my own personal experience yielded a framework that allowed me to think about multiple types of technology whether it be software, hardware, in the realm of computers, or even in the realm of manufacturing.

CHESS™ is an acronym which stands for multiple steps in the CHESS™ Framework. You will see the same acronym used for these multiple steps making it easy to remember one word throughout the entire framework. Like the actual game of chess, CHESS™ is meant to be strategic but also a framework that allows you to think a few steps in advance to position your company as the best in a competitive marketplace.

The CHESS™ Framework is a way to determine if you have the correct components in place to achieve your business goals. I consider achieving business goals as the definition of business success within the framework. Business success is made up of four specific variables that can be determined based on an assessment that I created through academic research. In this way, your business success can be deterministically forecast to increase the probability that you achieve success.

THE FORMULA

The CHESS™ formula is comprised of four variables that make up the output, which is your business success. The four variables are additive. In my research, I found that each variable about equally adds the same value to the probability of business success. As a result, all the variables inside the CHESS™

Framework are equal. If you fail to evaluate one of the variables, then it has an equal likelihood that you fail to achieve your business success. Put in other words, all four variables must be present and correct for your business to succeed in using technology to achieve your business goals.

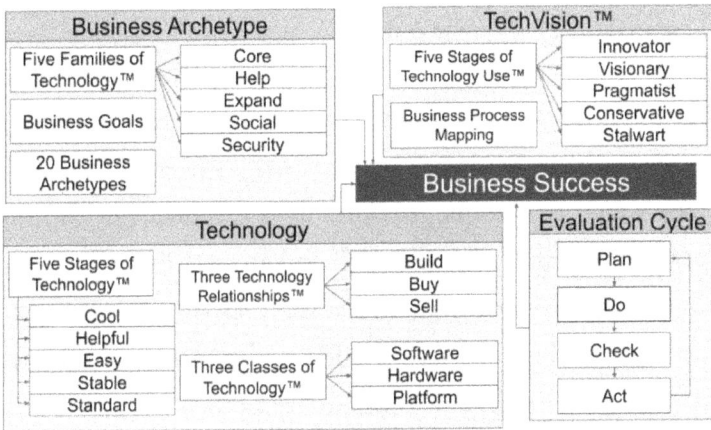

Business Archetype

Five Families of Technology™	Core
	Help
	Expand
Business Goals	Social
	Security
20 Business Archetypes	

TechVision™

Five Stages of Technology Use™	Innovator
	Visionary
	Pragmatist
Business Process Mapping	Conservative
	Stalwart

Business Success

Technology

Five Stages of Technology™	Three Technology Relationships™	Build
		Buy
Cool		Sell
Helpful		
Easy	Three Classes of Technology™	Software
Stable		Hardware
Standard		Platform

Evaluation Cycle

| Plan |
| Do |
| Check |
| Act |

I want to make sure that you understand this framework and the components inside of it. Because the CHESS™ Framework is the foundation for this book, I will cover the high-level formula in this first chapter and discuss the component variables at an introductory level. The remainder of this book will delve deep into each of the variables and where they fit inside the framework and why they are in the framework in the first place. In section 2, we will delve deeper into the framework and how it can be applied to specific problems inside your business by dissecting the four variables of the framework in detail.

However, in this section I want to give you the proper context to understand the CHESS™ Framework itself before we begin using it to address business problems.

BUSINESS SUCCESS

The reason this framework exists is to increase the probability that you achieve business success. Now, this term 'business success' will be unique to you as an individual. In my experience, using this framework and in my research talking with business owners, I found that CHESS™ can be applied to many types of businesses with different business goals. Where some businesses may view profitability as their primary driver of business success, other businesses view profitability as a tool to be a force for good inside their local community. In my research, I found that not every business seeks to become the largest possible business. As a result, business success for you may be outstanding growth internationally. But it may also be creating a business that sustains your lifestyle.

Business success is the achievement of your business goals through technology. This definition can be applied to whatever your business goals are. In my research, I found that many businesses do not fully understand exactly what their business goal is.

Because business goals can be difficult to articulate for some business owners, I created an assessment which lists multiple business goals and allows a business owner to rank the importance of those business goals. Business success is always a result of trade-offs. Because you, as a business owner, cannot have everything that you need for your business at once, you must create a strategy given the constraints in which you operate your business.

As we go through the CHESS™ Framework, keep in mind that the framework will accelerate you towards your business success. Yet the framework will not define what success is for your business. Only by having a clear understanding of where you want to go will the framework be helpful in achieving that vision.

Take a moment and consider your business goals and what business success looks like for you. Most business advisors suggest that a business think at least three months in advance, generally up to a year in advance. Beyond a year, market forces and new entrants into the market make it difficult to accurately forecast where you want to be. One of my business mentors suggested that business success be outlined only in broad terms five or even 10 years from now.

To me, this meant that I wanted to help 1,000 business owners over the next decade. Yet this goal, while seemingly concrete, is actually incredibly nebulous. And this is by design. Because it is impossible to understand exactly what the market will look like in a decade, I have a north star metric that is a quantity but the "how" I achieve this business goal can change.

In this same manner, you should look at where your business could be in the next decade. What is your north star metric? Then take a much narrower view of the next 3 to 12 months and get tactical. What steps do you need to achieve for business success to be realized in the next 3 to 12 months?

With a view of where you want your business to be in the next year, the CHESS™ Framework can now be used to increase the probability that your business will be where you wanted to be in a year through the ideal use of technology.

BUSINESS ARCHETYPE

The first and arguably most important component of CHESS™ is your Business Archetype. Your Business Archetype is the unique combination of goals and outcomes that form the personality type of your business. Your Business Archetype will determine the ideal areas of your business in which to invest an additional technology to achieve your business goals. The Business Archetype is one of the things that makes your business unique from your competitors.

Your Business Archetype is derived from the goals that are most important to your business in the next 12 months. This, then, is used to find the ideal two areas within your business that you should invest further in technology. While there are Five Families of Technology™ for your business, your Business Archetype highlights the top two which is your unique combination that allows you to compete when compared to other businesses.

Your Business Archetype also includes a component of your business differentiation. Whereby your overall business goals that you are striving for are the outcome of this framework, how you achieve those business goals through your differentiation strategy in the marketplace is a part of your Business Archetype. We will discuss the various competitive differentiation strategies later on in this book. But two companies with the same Business Archetype may approach their business success very differently based on the strategic business differentiation of the company.

The Business Archetype is mathematically proven based on an assessment that I created. The Business Archetype is valid for about a year. After this point in time, your business goals may change enough that your Business Archetype shifts. In fact, businesses will undergo a natural progression of archetypes depending on their goals for the next 12 months, the competitive environment in which they operate, and the macroeconomic climate. It is expected than that your Business Archetype will evolve as your business evolves.

TECHVISION™

Your TechVision™ attribute is the unique view that you as a business owner bring to your business and how you view of technology. This is the human element that operates your business processes. The TechVision™ element is made up of the Five Stages of Technology Use™ and business process mapping. These two components uniquely identify the way that your business can tactically use technology in the areas in which your Business Archetype has identified that you should invest.

The Five Stages of Technology Use™ recognizes that not everyone views technology the same way. For instance, some individuals are very excited about new technology and are willing to try new things. Yet other individuals wait to adopt new technology until they are certain of the benefits that technology will provide for the business. Along this axis, your business must recognize what it is likely to do based on your own personal views of technology.

It cannot be underestimated how important it is that you are comfortable, yourself, with technology in your business. Whereas a Business Archetype can tell you where to invest in technology, your TechVision™ will tell you how to choose a technology that will fit with your business.

Another component of your TechVision™ is the business process mapping process. The business process mapping within a business is an art form in the field of management. Whereas some management courses will teach a semester long course devoted specifically to the topic of business process mapping, much of the benefit of using this art form inside your own business can be unlocked in just a short time by reading this book. For this reason, we will cover business process mapping more in depth later on in this book. But suffice to say, it is really just drawing boxes and arrows.

Now there are some deep in the business process mapping field that will vehemently disagree with this simple classification of their art form. Indeed, for certain types of businesses, such as franchise owners, a detailed business process map is essential to replicate the process. If your business has multiple locations and you find yourself needing to write an operations manual, then the business process mapping inside your own company may be much more in depth then the simplistic view that other companies may need to use. But regardless of your size, using the basics of this management tool will allow you to more clearly understand how your business operates and provide a better level of service to your customers.

Combining business process mapping with the Five Stages of Technology Use™ yields your TechVision™. This attribute really speaks to how you as a human and your employees will interact with the technology that powers your business success.

EVALUATION CYCLE

No decision in business is ever perfect and, as a result, one must plan to evaluate and potentially reverse decisions as they are made. The Evaluation Cycle is a critical ongoing attribute in your business success through technology. Technology evolves over time. Your business also evolves over time. And, whether or not we like it, the market and our competition change too. As a result, the best technology from last decade may not be the best technology for this decade. But, how do we know when it's time to change?

The Evaluation Cycle is the answer to this question.

While there are many ways of evaluating things in a repeatable process, the one that I like is the plan-do- check-act or PDCA cycle. This lightweight framework allows you to quickly understand whether or not technology is fitting within your business to drive your business success.

We will cover each of these four steps later on in this book. The Evaluation Cycle should be undertaken on a regular basis.

We'll talk about how to regularly schedule lightweight evaluations of your technology. Through lightweight, regular intervals of testing whether the technology you have still works, you can avoid your business success being imperiled by choosing the wrong technology for too long of the time.

This Evaluation Cycle is easy to describe but can sometimes be hard to implement as there is a component of evaluation that may require expert opinion outside of your own, as a business owner. We will talk more about how to use this powerful cycle later on in this book.

TECHNOLOGY

Lastly, the final attribute in using technology for your business success is the technology itself. Yet, while this attribute contains within it the variable in the framework, there are also a number of steps inside this particular attribute that expand it into its own framework as well. The technology you are evaluating for your business will land in this attribute. Yet, there are many ways of looking at the technology that change this attribute itself and helps to best align with your business success.

There are really four components of this attribute. The first is, of course, the technology you are evaluating. This could be the brand name, the generic technology, or even an idea for technology that you want to build yourself. But there are three other components of this attribute that we will discuss in this book. These are the Five Stages of Technology™, the Three Technology Relationships™, and the Three Classes of Technology™.

The Five Stages of Technology™ deals with the notion of maturity. Maturity is one of those words used in technology to describe how complete and unchanging a technology is. A very immature technology is one that has not yet been proven. For instance, this may be cutting edge technology that is only been proven in a research lab at a university. Yet, a very mature technology may be one that has been around for years decades or even centuries. An example of the very mature technology is the wheel. While a round object no longer captures our fancy, as it once did the entire human race, this technology is used everywhere and is considered to be a foundational building piece in mechanics.

In a similar way, we can evaluate technology using a standard scale so we can talk about the differences in technical maturity.

Any technology you are evaluating also comes with the Three Technology Relationships™. These Three Technology Relationships™ are the ways in which your business relates to the technology in buying, building or selling. Often, when a technology does not fit a business, the Three Technology Relationships™ framework can be used to understand why the technology is not working. Each of these relationships carries with it a fundamental difference in your company's culture, as well as, tactical operational steps that you must do to have a proper relationship with technology.

The last component of this attribute is the Three Classes of Technology™. These Three Classes of Technology™ are the ways in which technology can manifest itself relative to the Five

Families of Technology™ inside your business. There are only Three Classes of Technology™ and every technology can fit into one of these three classes. Because these Three Classes of Technology™ are so foundational, it is important to quickly understand which class of technology the particular technology you are looking at falls under. Be warned that using the wrong class of technology, even if the technology is right, can cause your business to fail to meet its goals. The Three Classes of Technology™ provide clarity when you are evaluating whether or not to use technology inside your business.

Chapter 2

DISCOVERING THE CHESS™ FRAMEWORK

I sat there at our dining room table, a small table only big enough for two people, on the chair with a wobbly leg looking at my laptop, my phone in my hand. The mobile application that I had paid thousands of dollars was open on the screen.

It was wrong. The screen didn't look anything like what was in my mind. It didn't work. The bill was thousands of dollars over budget. My business was falling apart.

Though my technology friends had told me there was nothing wrong with my technology, my business had failed. Unpaid bills tallied up to thousands of dollars. No sales were coming in. Yet we had a technology environment that rivaled the best tech firms. We did everything by the book, the technology book.

As I faced a mountain of debt and a failed business I wondered, "where did it go wrong?"

This was the beginning of my quest to understand how technology can unlock the business of your dreams and how it can catapult your company into a realm you had never dreamed of.

MY STORY

Like many kids growing up in the 90s, I came of age in a generation where technology grew up with us. We were not defined by the Internet. I still remember when my family got our first Internet connection. I would spend hours online, learning how to code websites, interacting with my peers, and even starting a small eBay business.

I was too young to fully appreciate the revolution going on around me. But, as technology changed our lives, I understood that the books in the library were going to be replaced by a global network of knowledge that we could access immediately.

I grew up in a world where it was risky to put your credit card into a website online. Looking back now, from this initial interaction with the Internet, it seems almost silly how we had apprehension about technology that, now, has fundamentally changed our lives. People spend every day online, on websites, without thinking twice about putting in their critical information or transacting with companies around the globe. Technology has fundamentally changed how business operates. And by using the proper technology, a business can compete globally, instantaneously, and with companies much bigger than itself.

From a young age, I knew that I was going to be using technology in business. I was enthralled with the idea of doing more and being more than just one person running a small business. In high school, I attempted to run an online store through eBay. I tried my hand at drop-shipping and luckily broke even on my initial investment. I wanted to grow a business using technology because technology was cool. Technology makes it possible to do things, as one person in one firm, that were not possible when I was much younger. Technology makes it possible for us to live our ideal life and work on our dream business without needing the people, the capital, or the processes of the old way of doing business. Through a properly run business using technology, your dream lifestyle is unlocked.

I tried many different ways to make technology work in business before I finally found CHESS™. Some of the ways that I tried before included using way too much technology. For a while, I built the technologically perfect business, but never ensured that I had product market fit. And as I looked at those around me building their businesses, I found small business owners, who were lured away by the promise of the latest flashy technology, only to end up spending tens of thousands of dollars [or more] wasted and their business worse off or insolvent because of it.

Like you, I witnessed the rise of self-service tools for the small business owner. What was once as simple as having a website, quickly became a new platform or system that one had to figure out. The simple static brochure website gave way to a dynamic website that needed a shopping cart and e-commerce capabilities. A visually unappealing website gave way to complex user research and user experience studies that provided the best website possible, which when optimized to make customers spend more money.

I grew up in an era where technology and the Internet moved from the toys of the tinkerer to the tools of the entrepreneur to the formulas of the major tech firm.

It seems truer now today than ever before that the plethora of tools and options available to the small business owner has created an analysis paralysis whereby it is no longer possible to make a simple decision. Individuals are stuck learning the tools and optimizing systems so complex that they forget the need for the system in the first place. Small business owners forget that technology must serve their business.

MY JOURNEY

When I was consulting for a small business up in Alaska, I was struck by the reality of the complexities of using technology inside of a small business today. Whereas one used to simply type up a document in Microsoft Word, there are now at least 10, if not more, viable tools to use for something simple as word processing. Yet, adopting an online word processing vendor requires you to adopt their entire platform which means that you must now change how your business operates.

As I met with a small business owner on a weekly basis, I begin to realize that they had more questions than I had answers. I knew that there was a right way the technology could solve their problems in a way that technology could catapult their business and allow them to live their dream of running their small business. But I didn't yet have the framework nor the understanding to make that possible.

And the major technology vendors and even the managed service providers for small businesses were not helping. You see, most technology providers to small business are focused on getting you to buy a piece of software or hardware. I learned this firsthand when I was working to grow a consulting company and my vendors gave me sales quotas to push their software solutions on small businesses. These software solutions were not always useful nor beneficial to the small business owner. I could not in good conscience ethically sell these unnecessary solutions to small business owners. For instance, one particular technology provider wanted me to sell enterprise grade cybersecurity solutions to a small one or two person business. A small business doesn't need enterprise grade cybersecurity solutions. I knew this and yet, I was being forced to sell a product to meet quota.

I realized then and there that the problem with the technology industry, the reason that small business owners were not able to achieve the business of their dreams, was because the technology industry had become too complex. Instead of providing a solution, the technology industry provided more problems than their proprietary tools could solve.

Now don't get me wrong. Technology is not evil nor does the technology industry inherently want to be a villain. In fact, what is happening is that in the absence of clear guidance from customers, engineers will build the optimally configurable solution into their technology. What this means is instead of making the tough decision that you cannot do something, an engineer would rather provide you the option to configure the system to do it. As a result, the system grows in complexity. And what began as an altruistic effort to provide options to small business owners quickly becomes a configuration nightmare the small business owner must pay to understand, must pay to have implemented, and must pay to have maintained.

Against this unintentional complexity and confusing technology landscape, I began the journey of providing a clear framework to elevate the business needs and provide the proper place for a technology solution inside of a small business such that the small business owner was able to have the business of their dreams.

I realize that if I failed on this journey and I did not find a way of thinking about the technology issues in a small business that made sense to the small business owner who is not proficient in technology, such as the business industry, then small businesses would suffer and be unable to compete against larger enterprise.

A NEW OPPORTUNITY

I started my journey in academia. I've always been a fan of structured learning because I feel that a strong basis of theoretical and practical knowledge provides the best possible platform from which to grow and contribute.

As I worked through my Master's Degree in Cybersecurity, I began to realize that cybersecurity was a perfect embodiment of the complexities of technology that the small business owner faced through no fault of their own. Cybersecurity is a very complex, technically challenging, difficult field of emerging technology that the small business owner had to understand or face the chance that their business would no longer operate after a successful cyberattack.

Guided by my pragmatic view of how technology should fit with a business and faced with the complexity of a cybersecurity solution, I devised a new industry standard cybersecurity framework. This framework took the optimal components of the government framework for cybersecurity and applied it to the relevant points of a small business to ensure that it was safe.

This was the first epiphany I had on the journey to aligning technology with the business of your dreams.

My journey continued and I enrolled in my Doctorate program. For the first time in my life, I had a chance to apply my knowledge to a real-world problem and create something new. Grounded in theory stretching back over 80 years, throughout the technology revolution, I was able to draw out the challenges and potential solutions that small business owners were facing today in business.

After reading hundreds and hundreds of academic papers, talking with small business owners, and writing a 100-page dissertation myself, I finally had the beginnings of a framework that would become CHESS™.

When I looked at my dissertation and my hundreds of pages of academic knowledge and a decade of industry research and experience, I realized that no small business owner had the time to go through the same journey I had to understand what I had learned.

I needed to create a new opportunity for the small business owner to gain the benefit of my research, my experience, and my knowledge but do so in a way they could have it in minutes not decades.

THE CHESS™ FRAMEWORK

The CHESS™ Framework was born of the struggle to place technology back in its proper place of accelerating your business so you can achieve the dream business that you want. The CHESS™ Framework combines decades of academic knowledge, over a decade of industry experience, and a fair amount of technical prowess into an easy to consume computerized assessment that gives you the starting points towards the business success that you want.

The perfect technology strategy covered in this book is the amalgamation of my research, academic scholarship, scientific inquiry, and industry experience that provides you with the best answer to whether or not your business goals will succeed. You will learn that there are many components of CHESS™. While there are four main attributes of CHESS™, each of these attributes contains within it a host of knowledge and other frameworks that can be used and must be used appropriately in order for your business goals to succeed.

So, does it work?

Yes.

The CHESS™ Framework has been proven to work both academically, statistically, and in industry. These are three key distinct ways that we can arrive at the truth of this framework.

Now I know what you're thinking. Many industry gurus claim to have a framework that works. They claim to have the next big thing. They claim to have the truth.

What sets this framework apart is the care and precision with which I have arrived at the framework itself. I needed to know that this framework was true before I presented it.

I'll go into the academic validity of the framework more later on in this book. And that is because I can. Unlike a pop-culture framework, there is statistical validity and actual data that backs up the validity and veracity of this framework. It works. And not only does it work, it is academically proven to be the truth.

As we, as humans, currently understand how true knowledge is created, the CHESS™ Framework stands this test and passes with flying colors. The CHESS™ Framework is truth.

I'm glad to say that others have also achieved outstanding business results using this framework. Large companies have used the concepts contained inside this framework to turn themselves around and overcome adversity. Small businesses have outcompeted their major rivals and won in their market. Mom-and-pop shops have opened a second location. And individuals who simply wanted to live the life of their dreams with the business to support them have achieved these goals.

ACHIEVEMENT AND TRANSFORMATION

While my initial business from high school may have failed because I did not yet know about the CHESS™ Framework, I have seen many businesses since the creation of my CHESS™ Framework succeed. And I am proud to say that my own business now is based on the CHESS™ Framework and is succeeding. I was able to achieve the level of economic success that I wanted. I was able to achieve the lifestyle that I wanted.

But since the discovery of the CHESS™ Framework and the work I have done to apply it to small businesses, I still have a way to go on my journey.

As I look at the truth that I have discovered in the CHESS™ Framework and how it has been applied to businesses worldwide, with outstanding results, that allows them to live the business and life of their dreams, I am struck by a goal that I have not yet achieved.

When I left college, I lived alone for a time. During this time, I worked long hours at an unsatisfying job. Yet I worked with entrepreneurs to raise money for their business. I got to see dreams for other business owners fulfilled as I dreamed of having my own small business again one day.

It was during those days that I made a vow to help 1,000 small businesses achieve success and provide the owners with a lifestyle they had only previously dreamed about.

While I have the external markers of success and now have a successful business, I am driven by that initial desire to help small business owners. I am driven to help the next thousand small business owners use the CHESS™ Framework and the frameworks outlined in this book to find the perfect technology that unlocks the business of their dreams.

Chapter 3

WHAT CAN CHESS™ DO FOR MY BUSINESS?

Take a minute and envision your dream business. For some people, this may be working only a few hours a week. For others it may be multiple locations. And yet others may envision a company that they have either successfully sold or taken public. The dream business for every small business owner is different. Yet all of these dream businesses can be helped using CHESS™. The CHESS™ Framework is meant to be dream agnostic. Aligning the perfect technology in your business will unlock the business of your dreams.

When I was developing CHESS™, I was able to lean on small business owners of many different industries and sizes. I know for a fact that the challenges facing a business owner with 20 employees are very similar to the challenges facing a business owner just getting started today with no employees. This framework has been used in retail businesses, entertainment businesses, in professional services businesses, and in businesses with more than one location.

After the COVID-19 pandemic, the world woke up to realize that technology has fundamentally changed how business works. Whereas consumers used to be content with technology taking a secondary role in business, after the pandemic, customers now expect a virtual way to interact with your company. This pandemic has fundamentally changed how your business needs to align your technology strategy with your business strategy. This means that having the perfect technology in your business is one of the most important things to get right for your business success.

There are four key components of CHESS™. CHESS™ states that your business success is predicated on the alignment of your Business Archetype, TechVision™, Evaluation Cycle, and technology. Now we will go into what each of these building blocks are more in-depth later on in this book but suffice to say that all four of the uses must be aligned for your dream business to be unlocked. I talk a lot about the concept of the perfect technology unlocking the business of your dreams. This idea of alignment is really focused on making sure all these pieces are pulling together for the common goal and in a common method.

CHESS™ is one of the easiest ways to ascertain if your business is misaligned and provide you the steps and guidance you need to get your business back into alignment to grow your dream business. A quick example of misalignment here would be a company that seeks to grow but is not investing any money in advertising technology. The goal of this dream business is growth yet concrete steps are not being taken to achieve this dream business.

CHESS™ will also reveal deeper misalignment that may not even be recognizable to a technology expert. The common refrain, "technology does not work for my business" can actually be broken down into a number of specific reasons why technology is not currently working for your business. But most importantly, CHESS™ will also tell you how to change your use of technology in your business so that technology does work for you.

As a technology enthusiast myself I strongly believe that there is not a bad technology, but there is technology that is not ideal for your business. CHESS™ makes it possible to talk about technology in a way that clearly lets you express your desires, business goals, and thoughts about technology in a way that any technologist will understand and immediately connect with. CHESS™ provides you the language of technology in the language that you are most familiar with, the language of your business.

By using the CHESS™ Framework inside your business, your business will adopt technology that fits with your company, allows you to achieve business success as you define it, and unlock your dream business.

WHAT IS MY DREAM BUSINESS?

I talk a lot about the concept of a dream business. This is a very important concept for you as a business owner to understand. You see, I firmly believe that business strategy in small businesses is not actually at all beneficial. Now this is because there are multiple types of strategy and I'll get into these types in a second. But let's just say that it's best for you to have a dream "end state" for your business and then work towards that.

Many small business counselors suggest writing lengthy business plans and financial forecasts. I know that many grant programs also expect these lengthy documents to be produced by business owners. I actually think that this is an incredible waste of time for both the grant making agency and the small business owner.

You see, as a small business owner, you know your customers and your market. If you don't, then researching the market is the most effective way of gaining this critical context. But once you understand your market and you have a general idea of how you can provide value to your customers, it's best to sit back and imagine what a "future state" would look like rather than try to write down tens of pages and five-year financial forecasts. It's more important to go into business to figure out results than to write down pages of hypotheticals.

There are two types of strategy that a small business owner can follow. The first of this is causal strategy. The second is called effectuation. Both of these are valid types of strategy. Causal strategy is akin to being hungry and finding a recipe to cook something. It's highly likely that you don't have all the ingredients for the recipe. You must go to the store and purchase those ingredients. When you get back home, you then follow the recipe and create an excellent meal. You eat the meal, and you are no longer hungry.

Effectuation is akin to being hungry, going to the refrigerator and opening it up. Using the ingredients you have on hand, you make the best meal that you possibly can. You eat the meal. You are no longer hungry.

Both of these are valid strategies to feed yourself. In one, you may have a more elegant meal because you followed a recipe. Yet, when you look at your business, both of these strategies can be used to develop your dream business. The business plan is a recipe. Yet, you may not have enough money to buy all the ingredients to execute the business plan appropriately. Therefore, it is best if you keep an open mind and take opportunities as they arrive, not simply reject opportunities that don't strictly fit with what your view of what your business strategy should be.

This is why I talk about your dream business. Most of us have an overarching view of what our dream business is. This could be two locations. This could be making it to four years in business. This could be hiring 10 employees. Or this could even be going public on a major stock exchange. But this vision, this dream, will drive you to identify the opportunities that your business can take to best achieve this dream.

Using a combination of causal strategy and effectuation strategy allows you to move forward in achieving your dream business without spending too much time on the roadmap to get there. The reward is indeed in the journey for a small business owner, not necessarily the destination. CHESS™ allows you to quickly deal with opportunities that land in front of your business on your journey.

One of the key benefits of CHESS™ is that you can quickly identify both business and technology opportunities to be in alignment with your current dream business or not. But more than just identifying alignment, you have the ability to understand why an opportunity or technology for your business does or does not fit at this time.

HOW WILL CHESS™ HELP MY BUSINESS STRATEGY?

While CHESS™ was originally developed to identify the alignment between technology strategy and business strategy and help business owners adopt the best technology to achieve the business success, CHESS™ also helps small business owners every day to sift through the myriad of opportunities presented to them to find the ones that best align with their dream business.

One of the outcomes of going through CHESS™ is knowledge of your Business Archetype. This Business Archetype will change over time as your business grows and enters new markets, serving customers by bringing more value. Yet knowing your Business Archetype allows you to quickly identify business opportunities that are misaligned with your Business Archetype. Further, CHESS™ includes a component of TechVision™. Your TechVision™ also reflects how you think about new opportunities in the marketplace.

Combining your Business Archetype and your TechVision™ together allow you to look at an opportunity presented to your business and understand immediately if it is going to fit with your business culture and your business priorities.

One of the really cool things about CHESS™, as well, is that it includes a component of an Evaluation Cycle. This Evaluation Cycle gives you a mental framework to use when you are making decisions that don't have a certain outcome. An opportunity presented to your business may or may not benefit your business as you expect. It is an uncertain outcome. Yet by using CHESS™, you have a way of evaluating if the opportunity is still going to be beneficial to your business by the time it is completed.

CHESS™ is helpful for the small business owner to understand their business strategy and opportunities that arise during the course of business. Using CHESS™ allows you to move towards opportunities that help your business and help you unlock the business of your dreams.

HOW WILL CHESS™ HELP MY TECHNOLOGY STRATEGY?

CHESS™ was designed to be a framework to allow small business owners to identify the ideal technology for their business in such a way that minimizes the chance that technology does not fit your business culture or your business goals. At the heart of CHESS™ is the ability to identify if the technology is going to work for your business the way you think it will.

The four components of CHESS™ work together to help a business owner view a new technology through the correct lens. The Business Archetype carries the personality of the business into the discussion about technology. Your Business Archetype tells you what is important to you based on your business goals for that coming year. Immediate misalignment between technology and your Business Archetype will be obvious and will allow you to say no to new technology that represents wasted technology spending at this point in your business. That is not to say that the technology will not be a good fit later, but just right now, it represents a waste of your resources.

Your TechVision™ will showcase whether or not the technology is going to be a fit for your business culture. Each individual business has a unique culture that is derived from the business owner and the employees. This business culture carries with it certain norms and processes. If the technology is not going to fit your business culture, the TechVision™ component of CHESS™ will tell you this beforehand.

In the event that you are uncertain still about whether or not a technology is going to work for your business, CHESS™ includes an Evaluation Cycle. This allows you to quickly understand whether or not a technology will work for your business with the least money and time invested. In this way, CHESS™ allows you to minimize your losses in the event that you make the wrong decision.

The last component of CHESS™ is the technology and the way to think about technology. There are multiple components of the technology variable inside of CHESS™ that allow you to quickly understand and classify the technology you are looking at. This classification allows you to further identify whether or not the technology is going to work for your business or if there is a better technology that can solve the same problem.

CHESS™ really shines when it comes to helping your business pick the right technology to achieve your business success because business success looks different for everyone, CHESS™ is a flexible framework that will guide you towards the perfect technology to unlock the business of your dreams.

Let's dive into each of the four components of CHESS™ in the next section by addressing and overcoming the three most common problems small business owners like yourself have with technology.

SECTION TWO

Overview

This section of the book is the bulk of what this book is about. CHESS™ is a framework that I have spent over a decade developing and researching that provides you with the steps you need to unlock business of your dreams using perfect technology.

In this section, we will talk about how you can overcome common business challenges using CHESS™. It's important to note that these business challenges come from small business owners like yourself. I have talked with so many small business owners and all of them brought up one of these three types of challenges, if not all three of these challenges, that they face in their business.

As you read through this section, you will hopefully see problems that reflect the same challenges you have in your own business. There are two pieces of good news. The first is that you are not alone. Many other business owners have the same challenges and have used CHESS™ to find a way forward. The other piece of good news is that there are solutions to these challenges. Even if you are trying things today that you think might work, there are proven solutions to these challenges using CHESS™ that allows you to have the business of your dreams.

The three key challenges that we are going to discuss in this section are that technology does not fit my business goals, technology does not fit my business culture, and I don't know what I don't know. I'll go into each of these challenges in detail in each of the sections. But at a glance, technology not fitting your business goals is one of the most common problems where technology doesn't work for a business. Technology not fitting your business culture is best seen when you use technology but it just doesn't quite work the way you wanted to. And lastly, not knowing what you don't know is one of the most infuriating challenges as a master of your own business to be forced to admit that you don't even know where to begin getting enough information to make an informed decision.

Like I said, the good news is there are solutions to all of these challenges. Let's dive in and get some solutions for you.

Chapter 4

TECHNOLOGY DOES NOT FIT MY BUSINESS GOALS

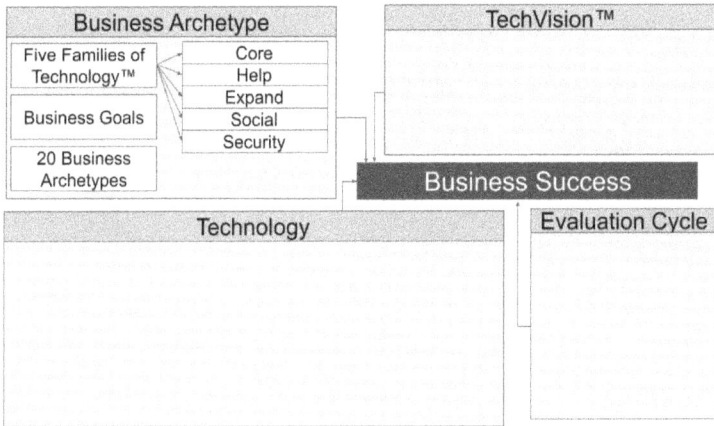

Business Archetype		TechVision™
Five Families of Technology™	Core	
	Help	
	Expand	
Business Goals	Social	
	Security	
20 Business Archetypes		**Business Success**
Technology		Evaluation Cycle

I t is very common to hear a business owner complain that technology doesn't work for their business. But yet, when asked deeper about what doesn't work with the technology and their business, most business owners struggle to adequately explain exactly what the problem is with the technology.

In most cases, I have found the actual problem is that the technology they are using does not fit their business goals. This requires a business owner to know their definition of business success. This is in itself a difficult task but one that is achievable for every small business owner.

When technology doesn't fit your business goals, it feels as if you are struggling against a process and outcome that leads you further and further away from the things that you care about. This is actually what is happening. Many times, technology producers purport that there will be benefits of using their technology that never materialize for you as a small business owner. This doesn't mean the technology maker was wrong. In fact, often that the technology was adopted for a purpose for which it is not ideal.

When your business goals and your technology align, you will wake up every morning having made progress towards your business goals without ever needing to work at that specific goal while you are away from your business. Indeed, it is possible to automate many of the tasks you are doing to achieve your business goals if you have the right technology.

Yet not every business wants full automation. This is why it is important for you as a business owner to define your business goals and then invest in the areas of those business goals.

That's right. Every business goal can be mapped back to one of five areas of technology. Now I'll get into what these Five

Families of Technology™ actually are in a second. In my academic and industry research, I have found statistically that every business goal fits to technology in one of these five areas. That is a scientifically proven mapping.

But the interesting learnings don't stop there. Did you know that every small business has an archetype?

A Business Archetype is the unique combination of the top two areas of technology investment that will unlock your business success.

Let's talk more about what a Business Archetype is and how it is comprised of the combination of your Five Families of Technology™, your business goals, and a description of each of the archetypes so you understand both what your business is, as well as, your potential competitors' Business Archetypes.

YOUR BUSINESS ARCHETYPE

Your Business Archetype is the unique combination of the Five Families of Technology™ and your business goals that differentiate your business from every other business in your industry. You may have heard the benefits of choosing a niche to better serve your customers. Your Business Archetype is reflective of your unique niche in how you approach and serve your customers.

Your Business Archetype may change over time. In fact, most small businesses undergo a shift in their Business Archetype from the early founding to their later years. It is not uncommon for your Business Archetype to shift every year or so as you refine what is business success to you as a business owner. If for one year you are focused on expansion, but in the next year are focused on survival, your Business Archetype will look very different between these two years.

It is important that you reevaluate your Business Archetype on at least a yearly basis. I recommend using a year in between Evaluation Cycles because it takes about a year to purchase, implement, and master any particular piece of technology. If you evaluate your Business Archetype sooner than one year, you run the risk of changing your area of technology focus such that you are unable to reap the benefits of technology in your business for your Business Archetype.

In my experience and in my research, I have found that the one year mark seems to map to the dynamic changes happening to a small business owner in their own business, their marketplace, their local business environment, and their technology investment. There's a lot that went into determining that one year marker, as you can clearly see.

The Business Archetype is made up of the Five Families of Technology™ and your business goals. The Business Archetype is a combination of the top 2 of the Five Families of

Technology™ based on your personalized assessment results. If you haven't yet taken your CHESS™ Assessment, I highly suggest you go to <u>chessprofile.com</u> and take your assessment now before continuing.

Chapter 5

FIVE FAMILIES OF TECHNOLOGY™

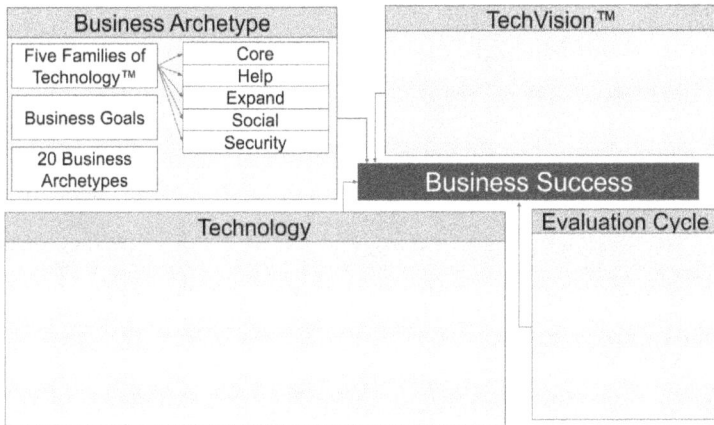

Business Archetype		TechVision™	
Five Families of Technology™	Core		
	Help		
	Expand		
Business Goals	Social		
	Security		
20 Business Archetypes		**Business Success**	
Technology		Evaluation Cycle	

Discovery Story

The Five Families of Technology™ was discovered in a watershed moment of my academic career. I had been working to classify the types of technology that small business owners needed to be successful. I had read hundreds of books and papers on the topic of small business owners and the technology they need to be successful. I had a mind map of tens of different categories that corresponded with classical business departments in large enterprises. I knew that small business owners needed marketing, manufacturing, customer service, and a central technology department. Yet, distilling a 50,000 person enterprise down to a one person startup kept me up at night, struggling to make an easy categorization system.

It was late at night in the winter up in Alaska. The sky was pitch black and the house dark. Everyone else had gone to sleep, except for me. Illuminated by the light behind my couch, I was reading about Michael Porter's value chain and how it was used in businesses. I saw the value chain graphic and suddenly realized this was the categorization system I had been searching for. Yet, Michael Porter's value chain is based heavily on a 1980s manufacturing company. It did not take into account the new technology-centric businesses that we have in this millennium.

Combining my research on small business technology, the value chain, and my knowledge of my own companies, the Five Families of Technology™ came to life that night. Based on academic theory that has been applied across millions of companies over 50 years, the Five Families of Technology™ distills the core areas that technology can impact in the modern business, regardless of whether it is one person working on a hobby or a 50,000 person enterprise spanning multiple countries.

From that one dark winter night, a new system was born that is now illuminating technology needs correctly in businesses around the globe.

Why is this concept important?

The Five Families of Technology™ is one of those fundamental business concepts that ensure your company achieves the business success you desire. In a world where technology continues to improve and the pace of business is ever quickening, adopting the right technology for the right reason at the right time in your business will make all the difference. The wrong technology could set you back and cause irreparable harm to your business. The right technology could catapult you ahead of your competitors and allow you to quickly change your business fortunes in days, not years.

Technology is all around us. There is no shortage of a new technology, new process, a new system that promises you the moon as a small business. Yet, too often, these technologies leave you simply poorer, having wasted time, and still seeking the promised results.

Technology is not easy. It has continued to get more and more complex. Looking back 100 years, technology in small business was simple. Now, to even accept credit cards at your company, getting paid for producing a product that you are the expert in, requires you to understand a myriad of intricacies of the technology that powers our payment systems. What's worse, your point-of-sale system can be, but is not always, tightly tied into your accounting records and your banking and payroll. Choosing the wrong point-of-sale system has cascading effects that impact every component of your business that needs money to function.

The Five Families of Technology™ provide a simple way of classifying technology inside your business that makes the core benefit of that technology to you as a small business owner instantly visible. Further, the Five Families of Technology™ allow you to map your ideal business success state back to the five areas where you should be investing in technology in your business. This direct link between the business outcome you desire and the Family of Technology to invest in has been proven, statistically, academically and in small businesses worldwide.

The power of the Five Families of Technology™ is immense. If you know what you want for your business in the next 12 months, the Five Families of Technology™ will direct you where to spend your money in technology within your business to achieve that business outcome in the next year. Yes, there are still complexities of technology that the Five Families of Technology™ do not address. But, the Five Families of Technology™ provide you the foundation to begin to untangle the web of technology solutions being presented to you every day as the small business owner. The Five Families of Technology™ allow you to classify each new opportunity, quickly understand if it will benefit your business, and then make a decision knowing you have enough information to be making an informed choice. And that is powerful for you as a small business owner.

There are five families in the Five Families of Technology™. These five families are Core, Help, Expand, Social, and Security. Combined, they form the acronym CHESS™. Just like CHESS™ is a strategic game, so to the Five Families of Technology™ allow you to be strategic in your business.

Let's dive into each of the Five Families of Technology™ separately. By understanding each of the five families, you will be able to apply the Five Families of Technology™ into your business immediately and unlock the benefits of having a new way of looking at technology as a small business owner.

CORE

The discovery story of this Family

Many years ago, I was leading a planning meeting for a small software development company. The company produced a mobile application that let users input their "to do" tasks for the day. One of the software developers, Jesse, was an incredible software developer and very experienced in this field. Jesse was making a point that we needed a better software development process to ship new features to our customers faster. As Jesse explained it, our customers paid for our software and paid for new features. Therefore, our customers directly wanted us to improve our software development process to give them new features faster. It was hard to argue with Jesse's logic amongst a bunch of engineers. Yet, I felt that Jesse's logic missed the true point of what our customers were actually paying us for.

I did what any good customer researcher would do. I went out and asked a couple of our customers what they were paying us for. The result was surprising. According to our customers, they were paying for us to keep track of what they needed to do and then check off when it was done.

According to our marketing department, our customers paid us to take control of their lives and live a more accomplished sense of themselves. According to our software development department, our customers paid for faster new features. Yet what our customers were actually paying us for was very simple. They wanted a place to write "to dos" and check them off when they were done.

From this realization, the Core Family of Technology was discovered.

Explanation of this Family

In business, every department and every function justifies their existence by linking their tasks back to what the customer needs. Yet too often, business departments stretch the impact that they have to a customer. Surely a better understanding of one's financial position allows the company to be more stable which allows it to take on more customers. In this way, the accounting department can justify what they do as impacting the customer. Yet behind this logic lies a key fallacy that many small business owners succumb to.

The Core Family of Technology deals with those technologies that directly impact what the customer is paying us for.

The key here is "directly impact" what the customer is paying us for. These are first-order effects. This is not an idealized end state or a lifestyle change. This is the boring answer of what the customer is paying us for when you strip away all the second order effects and marketing jargon.

The easiest way to understand the Core Family of Technology is through a few concrete examples that you as a small business owner will understand.

In manufacturing, the Core Family technology is the assembly line machinery that directly touches the end product that the customer will use. If you are a brewery, your core technology is anything that directly allows you to create beer. Putting the beer in a can is a core technology to your business. But monitoring your assembly line through an all-in-one platform, gathering data about your assembly process, is not Core Family of Technology. This is because your customer pays you for a can of beer in their hand. Not an experience, not a feeling of acceptance, and certainly not for you to have better insight into your manufacturing process.

If you are service provider, your Core Family of Technology is whatever you use to directly provide the service your customer wants. For instance, if you are an accountant, your customers paying you for their taxes and books to be done on time and correctly. Your Core technology would be any technology that allows you to produce the correct number for your customer. While you may market your services as helping your customers to save money on taxes or have a better understanding of their business operations, the core service your customer pays you for is simply the correct number in the correct location at the correct time.

If breaking down your company into these core services that your customer pays you for seems like reducing a really cool concept into something simple, that is because this. The Core Family of Technology is both unique to your business and highly constrained to only the direct process that takes raw materials and turns them into the finished product that your customer pays you for.

Where this Family is used in business

The Core Family is one of the most difficult of the Five Families of Technology™ to get correct. That is because of so much of what entrepreneurs are doing can be justified around providing a better experience or outcome for the customer. Yet maintaining the mental clarity to ask if the customer is affected directly through first-order effects or indirectly through second order effects is what separates technology belonging to the Core Family from technology belonging to the other families.

The technology in the Core Family of Technology is the fundamental technology that makes your business provide value to your customers. It is the direct technology that touches your raw materials and transforms them into a finished product.

Why is this Family important?

The Core Family of Technology is the single most important Family of Technology for your business' core processes. Optimizations in the Core Family of Technology allow you to better provide value for your customers. Likewise, adopting new technology in the Core Family of Technology allows you to bring on new products and potentially new product lines. The Core Family of Technology is so critical to your business that without it you would not be able to transform raw materials into your finished product.

Measures of implementation success

The Core Family of Technology is unique to your business. As a result, measures of successfully implementing Core Family technologies will be unique to your business. Expert guidance is needed to craft metrics of implementation success for the Core Family.

Yet, without expert guidance, it is possible to estimate the impact of Core Family of Technology through anecdotal stories. If your customers are saying that the quality of your product is increased, then the Core Family of Technology is performing well. If customers are concerned about varying quality in your final product, then the Core Family of Technology may be a place to investigate and adopt new technology. If your customers are complaining about long times between in order and receiving their finished good or service, then the Core Family of Technology may need to be improved to produce consistent output on a consistent schedule.

Tweaking the Core Family of Technology in your business is risky. Because this technology directly affects the quality of your final good or service, changes to this Family of Technology directly impact your customer loyalty, customer satisfaction, and your revenue. This particular family is a place where expert guidance is really needed to understand whether or not a technology will provide the intended benefit to your business and directly benefit your customers through a better finished product or service.

HELP

The discovery story of this Family

When I was researching technology for small businesses a number of years ago, I kept encountering a type of technology that did not seem to fit into any defined category. This technology did not directly change the quality of the product that a customer got. And it did not help a business get new customers. Yet, for some reason, businesses were shelling out thousands of dollars a month to have this type of technology in their business.

For example, a manufacturing company would purchase a machine learning system that would present better sales forecasts based on customer demand. Or another company had better employee satisfaction because they used an internal chat program that allowed everyone to stay in touch. And a food truck vendor was using software that let them track their supply chain vendors and reorder easily when they were running out of supplies like plastic forks and knives.

I struggled to understand why businesses would be purchasing technology like this as it did not directly allow them to charge more to their customers nor did it help them get more customers. And then I realized that this is a class of technology that operates *on the business* not necessarily in the core business.

Just as it is important to have a well-oiled machine to create products for your customers, it is equally important to have a regularly maintained machine to continue to function. Maintaining a machine requires all sorts of parts, a repair person, and a schedule for those repairs. Your customers do not care if you have the parts on hand to repair your machine. They only care that the machine continues to work. This meta-aspect of business is keeping the core value chain running properly.

Keeping the machine going involves a group of technologies called the Help Family of Technology.

Explanation of this Family

The Help Family of Technology is a group of technologies that make your business operate better. If you think of your business as a machine, the Help Family of Technology keeps that machine running. There are many types of Help Family technologies. This Family of Technology is particularly broad.

Some key areas of the Help Family of Technology are the ancillary functions around the core value of your business. For instance, having more data to make a better decision. Or more up-to-date accounting records and a better understanding your financial picture. Likewise, employee satisfaction and employee retention can be thought of as Help Family of Technology. Even something like Internet of Things devices to monitor a manufacturing line are a part of the Help Family of Technology.

The key difference between the Help Family of Technology and the other Families of Technology is that the Help Family of Technology does not directly impact anything involving your customers. The Help Family of Technology is full of second order effect technologies.

For instance, if you schedule appointments over the phone, scheduling software that only your employees use makes it easier to schedule an appointment. But, because no customers ever directly interact with the scheduling software to make an appointment, it is a Help Family of Technology because it makes your business operate better. No customer would pay more simply because your job is easier. But because your job is easier, you can serve more customers.

The Help Family of Technology is full of the second and third order effects. One can even envision software that makes it easier to manage devices which monitor your manufacturing line. This third order effect to the quality of your goods produced for customers is still within the Help Family of Technology.

Where this Family is used in business

The Help Family of Technology is found in your business any time you look at technology and think "it makes my business operate more efficiently" or "it makes my job easier". The Help Family of Technology is focused on you and your business, not your customers.

Some of the classic examples of the Help Family of Technology are the generic business functions. Every business has certain support functions that are the same or very similar across every type of business. If you look at a business school, most of the degrees conferred by a business school are focused on the ways to run a business.

As a result, the Help Family of Technology is often associated with these business school concepts. Things like accounting, legal, inventory management, people management, and general management are common across many types of businesses. Therefore, software that makes the jobs easier for these individuals falls squarely within the Help Family of Technology.

Why is this Family important?

The Help Family of Technology is a very important family for any business that seeks to operate beyond a hobby scale. Once the business becomes a separate entity from the founder, the number of people required to keep the machine moving and functioning appropriately increases. The Help Family of Technology shines in this realm and makes it possible to operate a business.

Without the Help Family of Technology, the entrepreneur would quickly find that their business breaks down as the core value chain comes under the natural stresses of running a business. Inventory would be delayed if not properly managed.

Lack of funding due to poor financial management would imperil the business. And no pipeline for additional employees would cause the business to fail when the last employee quits.

While this family seems to be wasted expense early on because it does not directly bring in new customers nor directly provide value to customers, this family becomes more and more critical as the business grows and reaches an age of stability.

Measures of implementation success

The Help Family of Technology can be implemented successfully with a focus on the benefit derived for the cost. The cost of this technology is easy to measure. Yet the benefit is murky. This is because of the second order effects of the Help Family of Technology.

For instance, does having better human resources software make your company more efficient? If properly measured, one could show that employee happiness increases. But does employee happiness increasing result in more output of your products? Does it result in more sales? This is why the Help Family of Technology is difficult to measure successfully.

As a result, some businesses seek to minimize the cost for the Help Family of Technology. By minimizing the cost for this family, they increase the risk that they do not see any benefit. Yet, because they do not measure the benefit directly, this family is thought of as a cost center and businesses seek to purchase the best technology for the cheapest cost.

For businesses that are able to measure the output of this family on their business success metrics, they find that the Help Family of Technology becomes a critical strategic advantage. For instance, a company that can consistently have a happier workforce will produce a better product and garner more customer loyalty. If you can measure the second order effect, then investing in a proper human resources system seems to be a prudent business decision that increases your competitiveness against another individual company in this industry.

EXPAND

The discovery story of this Family

Think back to when you were first starting your business. Put yourself back in the shoes of that new entrepreneur with an idea and a vision for what your life could be like when your business succeeded. You spent all this time getting your product ready and now you needed to find people to talk to and convert into customers. With the first few customers under your belt and your experience in delivering your product growing, you got hooked on this idea of making a living and making a difference. You set a vision for what your business success was.

If you remember this feeling as a small business owner, you know that the first thing you need to do after developing a product is find someone to buy it. Right after the Core Family of Technology, the Expand Family of Technology is likely to be your next investment as a small business owner.

As I researched small business technology, I found there was no shortage of individuals providing products and services in the Expand Family of Technology. This family is focused on growing your business. And growing your business is an exciting thing to do.

Explanation of this Family

The Expand Family of Technology is any technology that is directly focused on bringing in new customers or expanding existing product line to serve new customers. The Expand Family is your ticket to generate more revenue and achieve your vision of business success.

The Expand Family focuses on two major prongs for most small businesses, but it can also include esoteric ways of expanding your topline revenue. The first two easy ways of expanding one's business are to get more customers or to sell more products to the same customers. You can think of this as growing your business through reaching more of your initial target market or finding a new product that your target market currently needs. Yet, there are esoteric ways of growing your business as well. One of these would be mergers and acquisitions. The Expand Family can also deal with raising capital or purchasing another business as long as it is in the service of growing your revenue.

The Expand Family can be a bit like a casino patron sitting at the roulette table. Each spin of the wheel brings that the potential for a positive return on your initial bet. The difficulty with the Expand Family is measuring that positive outcome successfully. Just like gambling can become an addiction, so too can attempting to grow one's business and continuing to throw good money towards a technology area that will not return the results expected.

Where this Family is used in business

The Expand Family of Technology is used in business in activities that directly lead to new customers. Once again, if the technology is meant to optimize your monetary spending within the Expand Family, then it is the Help Family. Second order effects that get new customers belong to the Help Family. Direct new customer acquisition through technology is the Expand Family.

Examples of technology in the small business space that falls squarely in the Expand Family are your online marketing technologies, technologies for appointment setting, technologies that remind customers to show up or confirm actions, technologies that reengage customers after a sale has been made, technologies that allow you to enter new markets, technologies that allow you to find and acquire competitors, and technologies that grow your lead lists.

The technologies that are involved in your sales process almost all fall under the Expand Family of Technology. Defining your target market also falls under the Expand Family of Technology.

Why is this Family important?

Outside of the Core Family of Technology, the Expand Family of Technology may be the second most important for most new small businesses. This is because once a product can be made, it must be sold. New companies often struggle with the Expand Family because they do not yet understand the repeatable processes necessary to bring in new customers.

The Expand Family is important up until the small business has a new customer acquisition processes in place that sustainably and repeatedly provide new customers to their business. At that point, most businesses begin to focus on other families of the Five Families of Technology™ because they are no longer in a feast or famine mode.

Measures of implementation success

The Expand Family is focused on return on investment. The overarching metric for any technology in the Expand Family is a positive return on investment (ROI) as realized in topline revenue growth. Yet there are drawbacks with this measurement.

Measuring customer lifetime value is a complex task for many businesses. In fact, measuring customer lifetime value incorrectly means that a business will not invest enough money into the customer relationship to unlock all the potential revenue that could be obtained from the customer.

Yet the difficulty in measuring ROI often leads business owners down a dark path of expansion addiction. Business owners seek easy quantifiable metrics that show momentum. A false measurement of the Expand Family of Technology are second order metrics that could lead to new customers. For instance, one of the key examples of this is social media interaction metrics (e.g. likes, views, votes, etc.) that do not directly result in a new lead in a sales pipeline.

Likewise, an abundance of unqualified leads also points to wasted technology spending in the Expand Family of Technology.

Measuring the investment in this family must be done with a clear view of the probability of converting a pipeline lead into a customer. You need enough data to obtain a valid sales funnel with conversion percentages at each step in the funnel. Failure to achieve clarity around one's sales process leads small business owners to waste money in the Expand Family of Technology, such as continually buying lotto tickets on the hopes that they will convert into that elusive win.

Measured properly, the Expand Family of Technology is the first family that any business owner must master once they have mastered the Core Family of Technology with a repeatable process for delivering their product or service. Mastering the Expand Family and properly measuring the metrics is the key that unlocks your engine for growth allowing you to build a sustainable business that produces more revenue than you spend on the Expand Family of Technology.

SOCIAL

The discovery story of this Family

Think back to your time in the education system. Every test was individual. You didn't get to work with your friends to complete a test. You weren't graded as a group on your knowledge of history or mathematics or chemistry. Yet, when you became a business owner, suddenly you had to deal with suppliers and resellers, manufacturing and service processes that you may not have known about. Suddenly, you had more questions than answers and questions about where to even begin your search for those answers.

As any small business owner knows, building a business is a team sport. You need to be involved in your local community, knowing your local government to receive permits for business and expansion. As a service provider, you need to become a known name and a known brand in your local community to obtain clients. People do business with local businesses that they like and they think are giving back to the community.

You can even see this in the largest of brands. In the last few decades, brands have increasingly focused on how they give back to their communities. Whether it be through environmental initiatives like sponsoring and building parks or giving back through charitable foundations or even sponsoring local events in the town. Companies give back to the communities that support them. And companies need the support of the community to keep them flourishing.

This realization led to the discovery of the Social Family of Technology. Like everything else in business today, there is technology that helps your business be more plugged into your community, whether your community is local, online, nationwide, or even global. Technology allows your business to understand what your community needs from your company and how your company can help give back and strengthen the community in which operates.

The Social Family of Technology is distinct from your core value chain. For instance, consider a large company like Johnson & Johnson which is a large consumer focused healthcare company. Their core value is to provide healthcare products. Yet they also have a charitable foundation that improves the lives of millions in developing countries. They manage a large grant program to improve the lives of millions.

The technology that their foundation arm uses is directly related to the Social Family of Technology.

Explanation of this Family

The Social Family of Technology is any technology primarily used by your business to connect into your business community. These may be technologies for networking or social media technologies whose primary purpose is not to get you new customers. The Social Family of Technology is not just related to online platforms or social media. The Social Family of Technology can also be used for keeping track of events in your local community or organizing your company's response to needs that you identify that are not directly related to your core value chain.

This Family of Technology will depend on how your business seeks to interact with its business community. This may be as simple as discovering your business community on a local, national or global scale. It may be leading a business community on one of these scales. Or it may be simply keeping tabs on and talking with other business owners to exchange stories and advice.

Where this Family is used in business

The Social Family of Technology is used in business in one of two key areas. The first of these is establishing a presence in the business community. The second of these is actively contributing or leading in the business community.

Establishing a presence in the business community is also the first step towards educating potential customers, vendors, and suppliers. If you are warming up a cold audience for sales, this could be considered a part of your engagement in educating the business community about your offerings. In this way, the Social Family of Technology leads into the Expand Family of Technology because it produces new customers through second-order effects.

Another way to use the Social Family of Technology inside your business is to be involved in the local business community as either active participant or a leader in that community. This may look like newsletter software, or social media software, or even marketing specifically focused on building your brand within the local community. Your Social Family of Technology could be meeting reminder technology or technology to keep track of participants inside your local government who you wish to speak with.

Why is this Family important?

The Social Family of Technology is the most advanced Family of Technology for any business owner. This is because all other Families of Technology should be at a relatively mature state prior to beginning your Social Family of Technology investment. The Social Family of Technology can be thought of as general brand building and should not be focused primarily in driving new customers. As a result, investments in this Family of Technology requires sufficient revenue and stability of those revenue to invest excess capital in beneficial activities that cannot be directly tied back to top line revenue improvement.

Ignoring this Family of Technology when your businesses is ready for it imperils your reputation with the business community, your local community, or even the national community in which your business operates. At some point in time, your business needs to control the narrative of its story and its brand. The Social Family of Technology allows you to build, own, and correct the narrative about your business by being involved in the discussions that are happening.

Measures of implementation success

The Social Family of Technology is difficult to measure through traditionally thought of metrics. Topline revenue is not the primary objective of the Social Family of Technology. As a result, your company must come up with success metrics that outline how you wish to be engaged in your business community. It is possible that you could use social media metrics such as followers or engagement numbers with your social media posts as a simple metric to gauge the success of this family. Yet, more advanced businesses could look at metrics such as press release reach or foundation involvement in charitable giving. Other companies may wish to influence government and measure their success in the adoption or defeat of certain legislative activities. Still other companies can look at the Social Family of Technology and measure their success by the real-world events and networks that they grow.

Because of the multiple facets of the business community in which your business operates in the many ways that you can be social, technology for the Social Family of Technology requires a business owner who understands the other facets of their business and has grown them and supported them to a level of maturity that makes being social possible.

SECURITY

The discovery story of this Family

Courtney was working on her laptop at a local coffee shop and creating a website for one of her clients. Like she always did on a Tuesday, she got her large coffee and sat down at a table in the back of the coffee shop, away from everyone else but still close enough to feel like a part of the rush and activity in the shop. She put on her headphones and got ready to design a website for this client who needed the website for the upcoming launch next week for their new line of products. Courtney was excited as she began to work, quickly becoming engrossed in her work. She didn't notice when the man walking by in a suit hit her laptop screen and the coffee fell out of his hand.

The large cappuccino splashed all over Courtney's keyboard and her screen went black. The client's website was saved to that now broken laptop. Courtney no longer had a functioning laptop. Yet, her client was still expecting to launch a new product next week. What could she do?

If this story sounds familiar, then you are one of many small business owners who would benefit from the Security Family of Technology.

The initial discovery the Security Family of Technology was focused on cybersecurity and protection from hackers. But as I talked to more and more small business owners, stories like Courtney's continued to show up. The manufacturing company whose conveyor belt broke and destroyed thousands of dollars of products. The tax professional whose computer broke in the middle of tax season. The social media executive whose account was hacked and client pages taken over. Or the business director subject to spam emails claiming to come from their bank needing them to verify information. Stories of disasters narrowly averted abound as examples of why the Security Family of Technology as needed in your business.

Explanation of this Family

The Security Family of Technology encompasses everything that keeps your business safe. When you think about a business as manufacturing machine, the Security Family of Technology is the insurance you have that if anything goes wrong you can continue to operate. In businesses, this is an amalgamation of things like backups, antivirus, your insurance policies and your contingency plans for when you can no longer operate in your primary offices.

The Security Family of Technology recognizes that not everything in business is going to happen along the ideal outcome path. In programming, we call this the happy path. The happy path is what you expect to happen when you expect it to happen and how you expect it to happen. Yet, as soon as you step off this path, the Security Family of Technology takes over. If your laptop is always going to turn on, then one day it does not, the Security Family of Technology is there to make sure you continue to operate as a business.

Where this Family is used in business

Most businesses have some set of contingency plans that they plan to use if everything hits the fan. In your business, this likely includes things like cybersecurity such as antivirus or antimalware, scanning your emails for spam, and backing up your files. Yet beyond just these pieces of technology, you can also begin to think about the Security Family of Technology as those technologies that you used to manage your other risks, such as insurance, warranty tracking, and where you write down your plans for what happens if your office gets flooded tomorrow.

Why is this Family important?

This family is the third Family of Technology that you will likely need to master once you have your core business operating and the basics of your customer acquisition pipeline figured out. Every small business owner will treat the Security Family of Technology differently depending on the risk profile. But ultimately, you cannot gamble with the livelihood of your employees simply because you feel that you are taken care of as the business owner. The Security Family of Technology will be used in your business to ensure that anything that life can throw at you does not result in your business and your employees suffering as a result.

Measures of implementation success

The Security Family is successfully implemented in your business when disaster strikes and your business is able to continue to operate. As a measure of success, the proof is when you need your business to continue to function, it does. Many small business owners will be uncomfortable playing this game of brinkmanship, not knowing that the business will succeed and survive when faced with a catastrophic event.

As a result, interim tests can be used to measure the success of the implementation of the Security Family of Technology. These tests are mental exercises that the business owner walks through to envision a scenario and their potential response. It can be going so far as to conduct a live exercise of playing out a scenario and enacting your Security Family of Technology to ensure your business can still operate. The interim steps of measuring success for the Security Family of Technology are possible and build confidence that the plan will operate as intended.

For most small business owners, the Security Family of Technology will be the most foreign and require the most expert knowledge to implement successfully. This is because the Security Family of Technology not only requires that you understand what can go right but also requires that you understand the process well enough and have enough experience to understand what can go wrong. This is where experts, who work with multiple companies and understand what can go wrong, can truly help the business owner understand how to measure success of this Family of Technology.

Chapter 6

BUSINESS GOALS AND DIFFERENTIATION

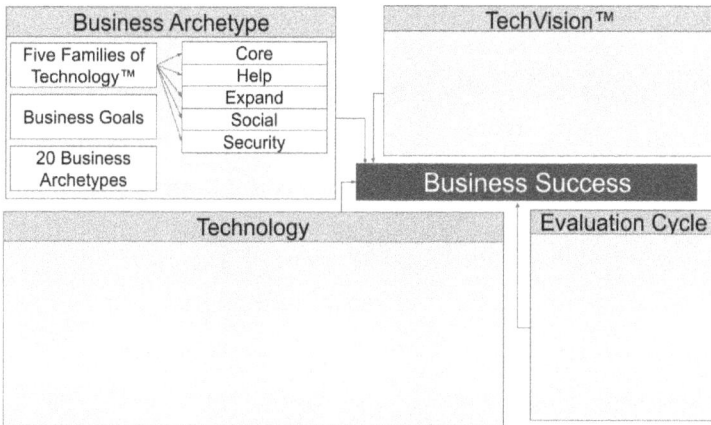

WHAT ARE BUSINESS GOALS?

Every day, your business is moving towards a goal. Whether or not you know what goal your business is moving towards and why depends your ability to set and execute on a future vision for your company. A business goal is a way of communicating both to yourself and to your stakeholders where you expect your business to be. There are a few guidelines to keep in mind when writing a business goal. We will cover these in this section, and you will end this chapter of the book knowing how to write a good business goal, how to measure your progress against that goal, and what to do when you achieve that goal.

There is no wrong business goal, except for not having a business goal. You as a business owner have an obligation to your business goals. But each business is unique. Your business goals may be large such as expanding into a new country. Your business goal could be simply to maintain the status quo in a turbulent market environment. Or your business goal could be unrelated to the financials of your company and instead be reflected in a social mission, such as a nonprofit seeking a business goal of societal change.

There is no wrong direction for a business goal. But you as a business owner must understand how to set a business goal that rallies your company and gets everyone excited to execute on your vision. There are a number of resources available online about how to set goals. The philosophy around writing future focused statements is the realm of business coaches and consultants. When I was teaching a course at University, I liked to keep the topic of writing business goals as simple as possible. This chapter is based on a part of the university course that I taught.

HOW DO YOU WRITE BUSINESS GOALS?

I like to write simple business goals. One of the acronyms that has been developed for writing business goals that I particularly like is the SMART framework. SMART stands for specific, measurable, achievable, realistic, and time bound. I like this framework because it is easy to remember and contains all the critical elements of the business goal that I have seen work well with multiple businesses.

Let's break down what each of these components are.

A specific business goal is one that is clear about what we are trying to do. A business goal of "grow the company" leaves much to interpret for the reader. Are we growing the company through new markets? Are we growing the company through new employees? Or growing the company through higher prices? This is not a specific business goal. A better goal is "grow the company by opening two stores."

A measurable business goal is one that we can attach quantitative metrics to. When it comes to measuring things in business, there are two ways of measuring. Quantitative is numbers based while qualitative is based more on words. We want a quantitative metric so we can understand our percentage progress. For instance, to "open two stores", we know that we need two new stores. One new store gets us halfway to our goal.

A business goal should be achievable. An achievable goal is one that we can do within our known limitations. If our company lacks financial resources, a business goal to open 10,000 new locations is likely not achievable no matter how hard we try. But a stretch goal is perfectly achievable. If we have two locations and we want to grow to five locations of the next year, that will be difficult, but it is not unrealistic if we are a restaurant with a large urban metropolis around us.

A business goal should be realistic. If we need a new leap in technology to accomplish this business goal and we are not a technology research and development company, then this goal may not be achievable. For instance, "increase human lifespan by 10 years" is likely not an achievable goal for a company in the event management space. If the goal does not align with our operations and purposes as a company, then it is likely not a realistic goal.

Lastly, a business goal should be time bound. This is because the element of time binds all of us together and limits what we can do. Time is a motivating factor. It makes the procrastinator work harder at the deadline and allows the planner to look ahead and see what a potential future should be. A business goal that is lacking the time element is simply a future wish. But adding the time element, for instance "in the next year", "in the next quarter", or "this week", adds a sense of urgency to the goal and allows us a point at which we can score our progress against that goal.

Combining these five elements together allows us to write business goals that are likely to be achieved or not achieved with a known reason why. Combining these five aspects of a SMART goal means we are now writing a goal that should invigorate the reader and provide them a clear understanding of what is expected. Instead of writing a business goal that is "get more revenue", we can now read a business goal that is "increase revenue by $100,000 through opening two new stores in the next six months." That is a business goal that instantly tells everyone who reads it what we expect to do, by when, and how we expect to achieve it.

HOW DO YOU MEASURE BUSINESS GOALS?

Measurement of business goals is of critical importance to the business owner. This is because a business goal begins to age as soon as it is written down. Perhaps the idea was to grow revenue, but the economy changed against us or a new law was passed rendering that part of our business unlikely to succeed in the near future. In this case, the business goal now is unlikely to be achieved. But how do we measure our progress made along the way? There are two frameworks that I like to use for measuring business goals. These are OKRs and KPIs. These stand for "objectives and key results" and "key performance indicators".

There are books written around how to write good objectives and key results (OKRs) so I will not go in depth in this book on material that is already fairly well covered in other business books. I suggest *Measure What Matters* by John Doerr if you want further reading. But at the high level, an objective and a key result is a pairing of your business goal and the steps you think you need to take to achieve that business goal. If the objective is growing revenue by opening two new stores, a key result would be opening the first store. Another key result would be opening the second store. It is likely that "opening a store" in itself has many key results. For instance, buying or leasing a space to open the store, ordering the store furniture and fixtures, and stocking the shelves with inventory. Key results are the steps that we need to take along the way to achieve our objectives. Key results can, themselves, have key results that need to be achieved to accomplish the objective.

Key performance indicators (KPIs) can be used with key results in objectives. For instance, if we know that we need to order five departments worth of inventory for our store, then we can track the order process for each of these five departments as a key performance indicator. This allows us to see when we have achieved our key result of ordering inventory which will achieve the objective of increasing revenue through opening stores. A key performance indicator is typically a number that can be shown on a report. When I was running a small company with just me and a salesperson, I created a quick Excel sheet that had a chart which showed the number of prospects we had reached out to, the number of leads that we had, the number of likely closed sales we had, and the dollar figure attached to the sales. While there is software to do this particular process, my Excel spreadsheet made it very simple for me and my salesperson every day to see what we likely had in future sales. I was then able to use this key performance indicator to forecast where we were likely to be the next month in our business.

With these two simple methods of measuring business goals, you as a business owner can now write a business goal that is achievable, break it down into components that need to be achieved to achieve the business goal, and measure your progress along those components. This allows you to move into the art of managing a small business instead of simply being a part of the process that executes the demands of your customers.

WHAT ARE THE DIFFERENT TYPES OF BUSINESS GOALS?

As I alluded to at the beginning of this section, a business goal is not always financial. There are six broad areas of business goals that I have identified for a business. These six areas are growth goals, financial goals, process goals, employee goals, external goals, and time-based goals. Let's cover each of these in turn.

What are growth goals?

A growth goal can be thought of as a goal that increases the business in some manner. This could be top line revenue. This could be number of locations. This could be the number of employees. Or this could even be a new product development and new product launch.

What are financial goals?

A financial goal is tied back to money. There are many ways to look at cash and cash flow inside of the business. The financial goal could be having more money in the bank. It could be raising a new round of capital. It could be increasing revenue or perhaps increasing the profitability of the business as a whole.

What are process goals?

A process goal seeks to improve the business efficiency. For instance, if a business owner identifies an inefficient process that could be changed through the use of new technology, adopting that technology and integrating it successfully into the business to change the process is a process goal. A process goal could also include something like decrease the number of defects and increase the quality of our products.

What are employee goals?

Employee goals focus on the most important part of your business, the humans inside of it. An employee goal could be increasing morale or increasing work life balance. Perhaps it is ensuring that your employees use all their paid time off for certain section of time. Or perhaps the goal is to decrease the number of sick days that your employees are taking. Employee goals focus on the career and well-being of the humans that help make the business function.

What are external goals?

External goals can also be thought of as the stakeholder goals for your business. This is where your business looks outside of itself and outside of its core business activities to determine if there are other aspects of society that it wants to improve. For instance, an external goal could be using only renewable energy inside your manufacturing plant. Or it could be decreasing the use of energy inside your business through the installation of solar panels. Or it could be changing the ways that you hire candidates to ensure you have a more diverse and inclusive work environment.

What are time-based goals?

The last type of goal is a time-based goal. Now, all goals should contain a time component. Yet the part that makes this type of goal unique is how we define time. For instance, a time-based goal could be very short-term, such as in the next week. Or could be a longer-term vision, such as "my five-year plan for world domination." Including the idea of a short, medium, or long-term goal as a type of goal is important because it allows us to understand what we need to do today versus what we should be planning for in the future.

As a small business owner, you now understand how to set a business goal for your company that is objective and time-based. When I was developing the CHESS™ Assessment for small businesses to understand where technology can help them accomplish their business goals, I included a specific time bound element. The CHESS™ Assessment looks ahead over the next 12 months and help the business owner to understand where their business goals fit and, given conflicting business goals, what is the most important to them. By using a generic statement of a business goal, I was able to build the CHESS™ Assessment that helps small businesses of many different industries to understand where and how technology can help them accomplish the business goals in the next year. Combining your business goals with the CHESS™ Assessment allows you as a small business owner to understand and have a better shot at achieving your own business goals for the next year. If you have not taken the CHESS™ Assessment, go to chessprofile.com to take it now.

Chapter 7

BUSINESS ARCHETYPES EXPLAINED

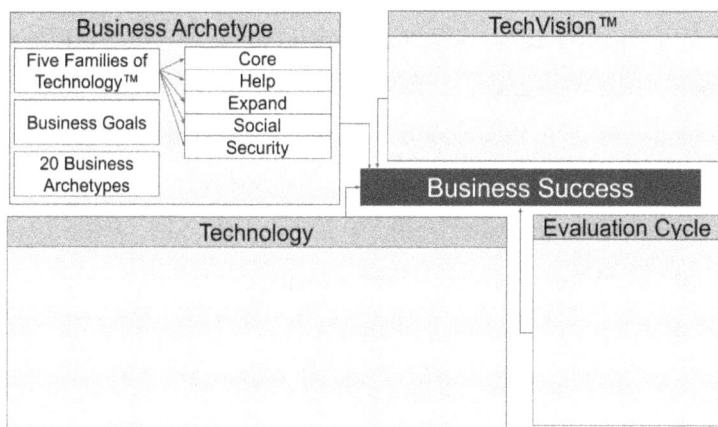

Business Archetype		TechVision™
Five Families of Technology™	Core	
	Help	
	Expand	
Business Goals	Social	
	Security	
20 Business Archetypes		
		Business Success
Technology		Evaluation Cycle

The CHESS™ Assessment will provide you with your Business Archetype, which is the unique two Families of Technology that correspond to the business success goals that you have over the next 12 months for your business. The Business Archetype is one part of your CHESS™ Profile for your business. This CHESS™ Profile for your business includes your Business Archetype as well as your TechVision™ score.

The Business Archetype is what makes your business unique and is likely to vary from your competitors. The Business Archetype also unites your business to other businesses in different industries that may make a product completely different from your business, but you and that business are aligned along your common goals and can learn from one another.

There are 20 unique Business Archetypes. Your business can only ever be one archetype at any given point in time. However, your archetype can and likely will change as your business grows, matures and as the macroeconomic cycle changes around you. For this reason, I highly suggest that you take a new CHESS™ Assessment every 12 months, at the most, to determine where your business is headed. This assessment should likely be taken every time a major event happens to your business. Some examples of major events happening to your business may include if the economy entered a recession, if you lose a key member of your staff, if you decide to enter a new market, or if your business grows by more than double in terms of either revenue or people.

Ultimately, the CHESS™ Profile must be applied to your business by you and your advisors. No single report takes the place of the decades of experience, training and education that an advisor will bring to your business and help you interpret your CHESS™ Profile. The Business Archetype is a component of your CHESS™ Profile and an advisor who is properly trained in my instrument will be able to help you understand what this means to your business.

At this point, if you have not yet taken your CHESS™ Assessment and gotten a CHESS™ Profile, I highly recommend that you go to chessprofile.com and begin the assessment now. It takes about 15 minutes and the results will allow this chapter to be more personal to you as a business. It will also allow you to understand which of these 20 Business Archetypes fits your business and begin placing the Business Archetypes of your competitors in your mind as you understand how to best compete against them by highlighting what makes your business unique.

With that disclaimer, let's begin discussing each of the 20 Business Archetypes based on the primary family of technology.

THE OPERATOR

Primary Family of Technology: Core
Secondary Family of Technology: Help

The Operator is primarily focused on their core business objectives with a secondary focus on their internal processes, or the Help Family. As a result, the Operator is primarily focused on operating. First, they operate *in* their business. Second, they operate *on* their business. The Operator is focused on what makes the customer happy and what makes the business run efficiently. These two focus areas combine to allow the

Operator deep insight into how and why their business must work to provide value to their customers and also run a well-oiled machine.

THE FACTORY

Primary Family of Technology: Core
Secondary Family of Technology: Expand

The Factory is primarily focused on their core operations with a secondary focus on expanding as a goal. The Factory focuses on what they do well and how they can replicate what they do well at scale. What the Factory does well is provide value to their customers. But what the Factory seeks to do after winning their customers is provide more value to more customers. This increased value may take the form of global expansion. It may take the form of an increased product line. Either way, the Factory is focused on what they do well and doing what they do well at immense scale.

THE PERFORMER

Primary Family of Technology: Core
Secondary Family of Technology: Social

The Performer has a flair about their business. They want to not only do their business very well, but they want to be impactful. The Performer provides value as their core mission to their customers. But then the Performer looks outside of just work. Perhaps they can be involved in a social cause. Perhaps they can be involved in the local community. The Performer always keeps one eye on what the customer needs and their second eye on how they can broadcast this to the greatest number of people for good. The Performer is intimately focused on running a good business. But a good business does not exist in a vacuum. The Performer knows this and takes joy in being a part of the business community around them.

THE GUARANTOR

Primary Family of Technology: Core
Secondary Family of Technology: Security

The Guarantor focuses on the business at hand and provides value to the customer at nearly any cost. The Guarantor's word is their promise to their customers. If you do business with a Guarantor, you will achieve the outcome you want. And you will be given it in any possible circumstance. The Guarantor looks to run a solid operation producing value. If there is a disruption in their business, the Guarantor has a backup plan. If there is a problem with their equipment, the Guarantor has an insurance policy ready to cover the cost. The Guarantor stands behind the value they provide to their customers.

THE ATTENTIVE

Primary Family of Technology: Help
Secondary Family of Technology: Core

The Attentive wants to run a good business that has the correct decisions and happy employees but also provides value to their customers. The Attentive will choose their business process over their customer. This is not a bad thing. In fact, the Attentive often has incredible customer loyalty and great internal processes. The business is likely to be a very good place to work. What the Attentive may lack in some customer delivery it more than makes up for in very good customer service. Customers of the Attentive know that the business owner is watching everything and paying attention at all times.

THE SYSTEM

Primary Family of Technology: Help
Secondary Family of Technology: Expand

The System small business owner is focused on process for the purpose of growth. This System has well defined internal processes. The System knows why their business operates the way it does. And knowing this, the System is now focused on expanding their business operations. It could be a new market. It could be a tenfold increase in customers. But the System knows that the business they have built will scale through proper internal processes, insight, and the ability to make the right decision at the right time to achieve this growth.

THE DEVELOPER

Primary Family of Technology: Help
Secondary Family of Technology: Social

The Developer is focused on the meta-picture surrounding their business. The Developer is highly attentive to the internal processes happening at the company. They understand how to tweak the machine that provides their customers value. But more than just their own machine, the Developer is focused on the industry and their environment as a whole. The Developer is likely to be consistently tweaking how their business works to maximize the employee productivity or happiness. But at the same time, the Developer is also bettering their overall industry, being a part of the local business group or perhaps lobbying for change at a national or international level. The Developer knows that constant experimentation yields results. And the experimentation must be done both inside one's business and in the surrounding business environment and society as a whole.

THE SHEPARD

Primary Family of Technology: Help
Secondary Family of Technology: Security

The Shepherd is a protector. The Shepherd is a protector of their resources, people, money, customers, and their reputation. The Shepherd archetype is focused on the business as a construct. This is the processes, people, and working environment. But the Shepherd is also acutely aware the downside risk that comes from operating a business. They seek to protect the business entity that they have made, ensuring that their customers and employees alike are protected in the event anything were to happen. The Shepherd probably has a plan B and even a plan C to make sure that their flock gets through turbulent times intact and well cared for.

THE FRANCHISOR

Primary Family of Technology: Expand
Secondary Family of Technology: Core

The Franchisor goes big from the beginning. The Franchisor has always had this idea of being the biggest in the market on their mind. Whether it's the most products or the largest number of customers, Franchisor wants to grow and build as their foundational DNA - but not just building for the sake of building. The Franchisor has operational excellence at the heart of what they do. They will provide the value to their customers and do so at scale. Grow first; provide value. That's the Franchisor model.

THE IMPROVER

Primary Family of Technology: Expand
Secondary Family of Technology: Help

The Improver is on a mission, a mission to change their industry and their business. The Improver always looks for new markets, new ways to provide additional value to more people. But right after getting new customers, the Improver looks at their own business to question how they can better provide for these new markets or these new products. How can we consistently grow and better ourselves and our business operations to sustain that pace of growth – this is the question that drives the Improver. The Improver is always looking to grow first and then better their business to sustain that growth.

THE PROMOTER

Primary Family of Technology: Expand
Secondary Family of Technology: Social

The Promoter likes to be in front of an audience; expanding one's business to new customers, acquiring new product lines, and having a bigger impact across the global stage is what the

Promoter wants. If they're not building their own business, the Promoter is involved with their local community. The Promoter advocates for change inside their market and out. They are the business that is likely to lead in social events, providing their name as a sponsor. Being a sponsor in social and the local business environment provides not only an opportunity for expansion and new customers but also fulfills their deep mission of being a part of something bigger than just their business and making a change there as well.

THE TACTICIAN

Primary Family of Technology: Expand
Secondary Family of Technology: Security

The Tactician is a cunning operator of a business that is methodical. This Tactician wants to run a larger enterprise. They want more customers. But for every foot of ground they gain in the marketplace, they look behind themselves and reinforce their position. The Tactician may appear to be moving slowly in their expansion, but they are gaining ground and winning that territory permanently. The Tactician thinks about every possible downside that their expansion will bring and prepares a response to it. If the Tactician is going to grow, they will maintain that growth and secure that growth at nearly any cost. The Tactician grows purposefully and safeguards himself against negative events that would threaten that hard-won growth.

THE MERCHANT

Primary Family of Technology: Social
Secondary Family of Technology: Core

The Merchant is gregarious, which is their true point of business. The Merchant knows what is going on in their business community, their political landscape, and the global economy. The Merchant always has one eye in the newspaper and understands what changes are coming before they happen. The Merchant is involved in rumor, the future, and what the winds of commerce are doing. But the Merchant does all this for the purpose of delivering customer value in repeatable fashion. The Merchant is always looking at the world around them for the purpose of making the sale and providing value to another customer.

THE BOOSTER

Primary Family of Technology: Social
Secondary Family of Technology: Help

The Booster is focused on the environment around their business and being a leading voice for their vision as to what the industry or the collective social good could be. But right after improving the business environment in which they operate, the Booster looks back at their own business and seeks to improve. If the Booster is advocating for better working conditions, they turn right around and implement these recommendations inside their own business. The Booster seeks to better their industry and lead by example, using their own enterprise as a test bed for the ideas they want everyone else to adopt.

THE NETWORKER

Primary Family of Technology: Social
Secondary Family of Technology: Expand

The Networker will focus on the broader environment for the purpose of growing their business. The Networker has a keen business sense of how the broader macroeconomic environment fits together. They understand the local business community, the political winds that may be coming their way, joint venture opportunities, or even a competitor on the ropes who may soon fail. But they keep all this information in their head for the purpose of finding new customers or new product lines. Behind everything that happens in the macroeconomic environment, the Networker is ready to find new business opportunity through new customers or new products.

THE GUIDE

Primary Family of Technology: Social
Secondary Family of Technology: Security

The Guide is focused on the broader environment in which their business operates but they have an understanding of where the pitfalls are in that industry. The Guide will seek to better the overall environment for everyone but also ensures their own business will not succumb to a negative event. The Guide may advocate for other businesses to adopt the security and protection measures their own business has already adopted. The Guide likely knows the best way to avoid the danger that businesses like theirs may face and they want to share this with every other business who can benefit.

THE SURVIVALIST

Primary Family of Technology: Security
Secondary Family of Technology: Core

The Survivalist is ready for anything. They have insurance, they have contingency plans, and they have proper cybersecurity. No matter what happens, they will be able to provide the value to their customers they have contractually obligated themselves to. The Survivalist will be the last business standing in their industry and still provide value to their customers. The Survivalist focuses on keeping the business alive first. After the business is guaranteed to survive, they will fulfill their obligations to their customers and provide the value that they signed their good name on the line to provide.

THE MONITOR

Primary Family of Technology: Security
Secondary Family of Technology: Help

The Monitor is focused on operating their business in a secure and efficient way. First, the Monitor ensures that every possible negative outcome has been considered and the proper mitigation plan put in place. After that, the Monitor looks at the way the business currently operates and seeks to better it. Are there inefficiencies that could be driven out? The Monitor will find these and devise a way to ensure these inefficiencies never happen again.

THE FORTIFIER

Primary Family of Technology: Security
Secondary Family of Technology: Expand

The Fortifier is focused on building a rock-solid fortress in new markets. Before even entering a new market, launching a new product, or acquiring a new customer, the Fortifier has already considered all the potential downsides and crafted a mitigation plan for nearly every risk. Plan before taking a first step, but always take the next step towards growth. The Fortifier expends company resources ensuring that any growth is protected against backsliding and the company will not be imperiled by growing. But grow the company must. And thus, the Fortifier builds a plan for every negative possibility and takes that step to increase revenue.

THE STALWART

Primary Family of Technology: Security
Secondary Family of Technology: Social

The Stalwart is the shining beacon of the best operational contingency plans in the industry. The Stalwart knows that they will survive, and their business will continue. From this knowledge and this strong position, the Stalwart advocates for economic change in their industry. If there is a new law to be passed requiring new standards, the Stalwart has implemented these standards themselves and knows that they work. Therefore, the industry should adopt them. The Stalwart does not change unnecessarily but seeks to change the industry around them. The Stalwart is focused first on maintaining their business through thick and thin and then ensuring that the industry itself is suitably and similarly prepared for any possible disaster.

These are the 20 unique Business Archetypes. Based on your CHESS™ Profile, you were assigned one of these Business

Archetypes. You should be able to identify other businesses like yours across many industries based on your shared Business Archetype. Your Business Archetype will change based on your business goals as your business grows and evolves. Your Business Archetype should be the starting point for evaluating any new opportunity that you encounter. By working with a certified fiduciary technology advisor, you can use your Business Archetype to identify the ideal technology for your business to unlock the business of your dreams.

Chapter 8

WRAPPING UP "TECHNOLOGY DOES NOT FIT MY BUSINESS GOALS"

A s we wrap up this section, I'd like to address some common concerns that business owners have shared with me over the years. These concerns relate specifically to the common problem that "technology does not fit my business goals." There is no particular order to these concerns. I wanted to take a bit of this book to address the concerns directly. This section might help you as well if you have some of these thoughts after reading the previous chapters.

I've tried lots of frameworks. What makes yours different?

The CHESS™ Framework is backed by decades of experience in the field and academic research. Not only is the CHESS™ Framework academically valid and statistically sound, but it is also generally applicable to multiple types of businesses. The CHESS™ Framework is simple to implement and provides an easy way of thinking about how technology affects your business and your chances of business success. Small business owners in many different industries have taken the CHESS™ Assessment and found that their Business Archetype is valid and useful to their pursuit of business success.

My company doesn't fit into these Five Families of Technology™.

This is actually a common objection that some business owners have. It is difficult to believe that something as unique as your individual business can be generally categorized into a framework that suggests there are only Five Families of Technology™ that you can invest in. But every business I have spoken with that has this objection and illustrates a "unique technology" still fits within the five areas outlined in the Five Families of Technology™. By asking a few pointed questions about the technology, it is clear where it fits inside the business and where it fits in the Five Families of Technology™. If you think you found a technology that doesn't fit within the Five Families of Technology™, I encourage you to reach out to me and together we will find it the proper home.

None of these Business Archetypes fit my business. My business is unique.

The CHESS™ Assessment will make it clear which of the Business Archetypes actually fits your business based on your business goals. Every business will fit into one of the Business Archetypes. Often a business may seem to fit into multiple archetypes based on the descriptions alone which makes it difficult for the owner to determine which archetype actually fits their business. But the CHESS™ Assessment provides for a comparative analysis that is statistically shown to resolve these edge cases and cleanly place your business in a Business Archetype. The Business Archetype where your business currently is may not be your ideal Business Archetype. But every business will transition through multiple Business Archetypes throughout its life span. It could just be that your ideal Business Archetype is coming once you address the measures of business success that you need to solve before you can reach your ideal state.

How can you reduce my complex business down to just one Business Archetype?

Through years of academic research and statistically valid studies, business goals were narrowed down and scientifically mapped to the Five Families of Technology™. From here, the Five Families of Technology™ were used to create the Business Archetypes. Further, these Business Archetypes were also scientifically shown to be valid and useful. Even a complex business ultimately has similar business goals as a simple business. Because of the similarity of what constitutes business success for your business to other businesses that have the same dimensions of success, it is possible to classify businesses in different industries who share common markers of what success means to them into the same Business Archetype.

I don't know what my business success looks like.

This is an objection that can really only be solved by the business owner and their advisory circle. Business success can mean multiple things to the individual business owner. Business success could be financial freedom for you and your employees. Business success could be brand recognition in every household. Business success could be a store in every town. Ultimately, business success is up to you and your advisors to define. This is because it is you and your advisors that will be responsible for bringing about the actions to achieve that business success.

How can I ignore the cool new technology that I see?

Oh, the new toy syndrome. Every new technology brings about exciting potential future benefits. Behind every new piece of technology is the future utopia that we have all dreamed of. Yet it is fairly easy to ignore the cool new technology when you realize that the probability of success of any new technology is incredibly low. Further, any new technology needs businesses to use it to work out the bugs. Do you really want your business to spend money in order for another company to learn how to improve their product? For some this may be an action that constitutes your business success. But for others, this warning will be enough to let the cool new technology become the standard technology before your business uses it.

I don't think technology can ever fit my business.

As I will cover later in this book, technology can be many different things. Technology does not necessarily mean high-technology, mobile phones, or the latest social network. If your business is manufacturing, technology may be a new assembly line. If your business is a service, technology may be as simple as a reminder for people to show up to appointments. Technology at its core takes a process that would be, otherwise, done by a human being and assigns a machine, whether hardware or software, to do that process instead. Your business can use technology and it will fit in your vision of business success.

I've already spent money in the wrong Families of Technology according to my Business Archetype. What now?

In business, there is a concept of the sunk cost fallacy. One of the hardest things, as a business owner, is to admit a mistake and move on. The sunk cost fallacy warns us against investing additional money in a decision that we now know was a mistake to attempt to salvage the mistake. Sometimes, business teaches us lessons through the school of hard knocks. And the best thing to do is admit we have made a mistake, stop any further losses, and change course to the desired state that we want to pursue instead.

All my competitors are investing in every new technology. Won't I be giving up market share to them?

It is dangerous to compare one's business to one's competitors. It is highly possible that you are competing with businesses that write off their new technology investments and have a much larger research and development budget than your business does. Likewise, there is no guarantee that adopting new technology is bringing your competitors new business. It may be the fact that the consistent thrashing between technologies is actually frustrating employees and driving away loyal customers. Do not be enticed by the latest technology and think that your business immediately has to adopt that new technology just because your competitors are.

My industry is really competitive. I can't wait 12 months for business success.

Some industries are actively being defined by their participants. Waiting a year to see the result of business success may be too long for your business. Yet, most businesses do not exist in an emerging market that is redefining a section of the business landscape. Even if your business is in a hypercompetitive industry, thinking in quarters and a year allows you to plan out what the future of your business and your industry may be. By being the stable business in an otherwise changing industry, you may garner more customer loyalty. But most importantly, 12 months is the period during which a technology should be evaluated. This is because many new technologies may seem to not fit the business initially but with training and repetition, become the backbone of your business process. Time is really your best friend for evaluating whether or not a technology will fit your business.

*****

A Business Archetype provides your business clarity. Could your business benefit from a defined path to achieve business success? Would you fall in love with your business again if it was successful? Would knowing your most important areas of technology investment help your business?

If you're like most small business owners, the answer to all of these questions is yes. And that makes sense because one of the hardest things as a small business owner is to know that you're taking the right steps.

Let me tell you about a company that was able to use its technology appropriately to achieve its business goals. This is a company in the financial services space. This particular company had about 30 employees and was focused on building financial services products for individual investors.

When I engaged with this company, they had difficulty knowing what their business goals were. Working with the company, I found that they were engaged heavily in expansion mode and needed to build a market for their products and services.

As a result, this company needed to focus on its Expand and Core Families. This company was a classic Franchisor Business Archetype. The company needed to create new products and fulfill the individual needs of its customers.

By investing heavily in technology that made it possible to repeatedly launch and test new products in the marketplace, they were able to achieve product market fit. This allowed the company to quickly scale up and serve more customers once their products were performing as they expected. By emphasizing the expansion of their marketplace through new products and new customers, they were able to dominate an emerging market.

This is a classic case of aligning technology inside the company with the desires of the business owners. Had the company spent time investing in the best accounting systems or integrating their payroll and accounting systems into a complex web of technology, they would have missed the market opportunity presented to them.

In fact, this particular company was able to avoid the traps of investing too much money in technology that did not align with their business success vision. For instance, the company was able to invest only the necessary funds in their security regime to ensure that their customers' and employees' data remained safe. Yet, once they had met the minimum, they made the conscious choice of stopping further investment to focus instead on finding a product that would serve their market.

Likewise, the company invested only enough in technology to meet their social obligations inside their industry. By limiting their investment in the social family to off-the-shelf software that they did not have to modify, the company was able to reap the benefits of being social within their industry without incurring additional technology costs in building a complex system that did not align with their ultimate business success.

Because the company focused on technology that allowed it to repeatedly produce high quality products and, through a marketing effort, get those products to market, they were able to win in their segment. The focus on technology to ship new products produced a process that was unique to this particular company. Likewise, their investments in core technology to fulfill the customer needs were proprietary to the needs of this new emerging market.

Because this company invested so heavily in technology that was aligned with their goals for business success, they were able to take the Franchisor Business Archetype and apply it successfully to their business.

If you are like most small business owners, this success story mirrors what you want for your business. It is not difficult to have this level of insight for your business as well. The CHESS™ Assessment will unlock the knowledge you need to apply your Business Archetype and business goals to your business success.

By aligning your Business Archetype in business goals with your business success, you will be able to find technology that fits your business goals.

If you have not yet taken the CHESS™ Assessment, go to chessprofile.com today and take the assessment now.

Chapter 9

TECHNOLOGY DOES NOT FIT MY BUSINESS CULTURE

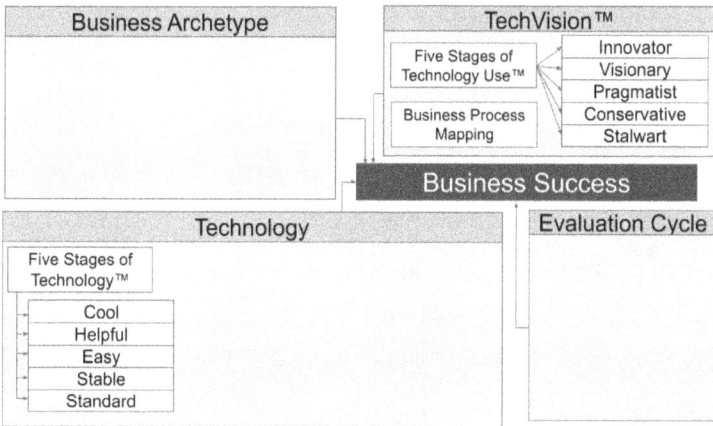

Business Archetype		TechVision™	
		Five Stages of Technology Use™	Innovator
			Visionary
			Pragmatist
		Business Process Mapping	Conservative
			Stalwart
		Business Success	

Technology		Evaluation Cycle
Five Stages of Technology™		
	Cool	
	Helpful	
	Easy	
	Stable	
	Standard	

In my work with small business owners, I have never heard a business owner express that technology does not fit their business culture. But often, what I'll hear is that technology doesn't fit their business or the technology is convoluted and difficult to use. This is actually an expression that the technology being adopted by this company doesn't fit the culture of the business. Very rarely is technology poorly constructed. Often, however, the technology does not fit the process of the business. This causes a cultural conflict between the process envisioned and codified within the technology and the actual processes being used by the business.

To delve into what a business culture is, as it pertains to technology, we will need to look at three distinct concepts within your business. The first of these is business process mapping. Next, we'll talk about the Five Stages of Technology Use™. Finally, we will touch on the Five Stages of Technology™ as well.

The first concept of business process mapping is a defined field that teaches business owners how to write down the abstract processes that occur in everyday business. Your business is just a set of processes that produce outcomes for clients. As we dig into business process mapping, I will explain a simple set of rules that you can use as a framework for beginning to understand your business more intimately.

The second concept we'll cover is the Five Stages of Technology Use™. The Five Stages of Technology Use™ is a concept that I develop based on my industry experience and academic literature. It takes advantage of existing research to help align your view of technology. Each individual business owner may have a different view of technology. This view of technology influences your starting position when you consider adopting technology for your business. We'll delve much more deeply into the theory and application of the Five Stages of Technology Use™ as it pertains to your CHESS™ Profile.

Lastly, we need to talk about technology itself. There are Five Stages of Technology™ that I identified based on my work with clients and industry experts. These Five Stages of Technology™ also form the acronym CHESS™. And it is an incredible framework for addressing technology before you begin adopting technology to meet your business goals. Even the right technology will not fit your business culture if it is in the wrong stage of the Five Stages of Technology™.

There's an old saying that "people don't quit companies, they quit bad culture". You, as a business owner, have worked hard to create a culture. This culture is the manifestation of how you expect your employees to act, how your employees interact with customers, and how you interact with your employees and your customers. Your business culture is very difficult to build because it must be built every day. Yet, nothing destroys a business culture that you have worked hard to build faster than a wholesale process change forced upon your company by new technology adoption.

Not every technology brings with it a process for that area of business that is compatible with your company. If you are used to providing much authority and trust to your employees, your employees will consider a system that has many permissions and security controls as the eroding the trust you have placed in them. If an employee used to be able to perform all the functions of their job themselves, but now has to ask you for permission to move past the particular step due to the technology you adopt for a business process, you have now signaled incredibly strongly to this employee that you do not trust nor have confidence in their skills.

Nothing erodes your business culture faster than technology that does not fit with the processes of your company. Because you worked so hard to create a positive culture, inadvertently signaling to your employees that you no longer trust them nor their skills, even if that is never your intention, will destroy business culture. A destroyed business culture takes with it your top performers and your top talent.

While some employee dissatisfaction with new technology is to be expected, you, as a business owner, must avoid turbulent effects of adopting new technology without the cultural considerations. What we will cover in the next few chapters provides a framework to think about adopting technology for your business culture to avoid employee turnover while increasing the efficiency of your business through technology. It is possible to unlock the business of your dreams through perfect technology while maintaining the strong business culture you have built.

Chapter 10

BUSINESS PROCESSES AND TECHNOLOGY

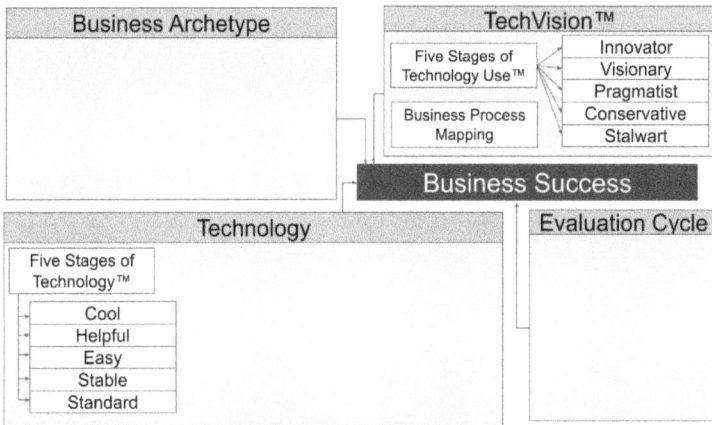

Business Archetype		TechVision™	
		Five Stages of Technology Use™	Innovator
			Visionary
			Pragmatist
		Business Process Mapping	Conservative
			Stalwart

Business Success

Technology	Evaluation Cycle

Five Stages of Technology™
Cool
Helpful
Easy
Stable
Standard

At its core, technology is used to make a process more efficient. As a result, every technology revolves around the concept of a process. A process is just a series of repeatable steps that take input and produce output. Your business is just a process that takes raw materials and produces the final product that your customers want. And your business process is itself made up of processes. Taking raw materials and unloading the truck and storing them in your warehouse is a repeatable process. Your monthly financial report review is a process that you do with your accounting team. But processes are notoriously difficult for small business owners to document because they lack the language to document a process.

When I was teaching at University, I had the good fortune of teaching a management information systems (MIS) class to undergraduates. I taught the first part of a three-part capstone series for these MIS students. My course was focused on equipping them with the tools to speak the business process language. While I'm not going to delve into a semester worth of a university course in this book, I am going to pick out the highlights of my course to help you as a small business owner to understand exactly how to document your business processes.

WHAT IS A PROCESS?

A business process has a start and end and a series of steps within these start and end parameters. There are many ways of explaining a business process, some of which actually require a university degree to fully comprehend. But my favorite way of explaining the process is with something simple – boxes.

If you were to think about the process of pouring a bowl of cereal for breakfast, it may look something like this: You get the milk from the fridge. You get a bowl from the cupboard. You get your cereal from the pantry. Let's call this get raw materials. Then you combine all the raw materials together. Note that it is very important that you put the cereal in the bowl first so you know how much milk to add. It is also important that you put the milk in the bowl not in the cereal box. We can call this combine raw materials. The process then ends with you eating this bowl of cereal. We can call this delivering the final goods. The image below shows both the specific process we talked about and the generic process.

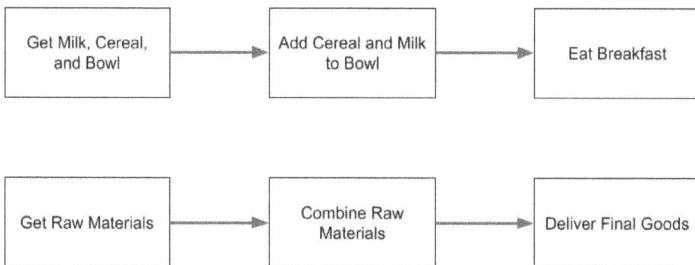

HOW DETAILED SHOULD A PROCESS BE?

Our previous example involving the process of making breakfast really shows the layers of a process that we can use. In the abstract, taking raw materials, combining the raw materials, and delivering a final product can explain the process for many types of companies. Yet moving a level deeper into the process, we discover that we were talking about making a bowl of cereal not building a new car. Even within the idea of making a bowl of cereal, at a large-scale we realize that there are processes around getting the cereal from the pantry and putting that cereal back in its correct spot on the pantry.

This realization of the multiple layers of the process means that we need to establish some rules about how we document these processes.

THE FOUR SIMPLE RULES FOR DOCUMENTING A PROCESS

There are really four key elements to document a process that every business owner should know. These are the idea of a flow from one process to another, typically shown as an arrow; the process steps, typically shown as a box; a decision point between two paths that your process can take, typically shown as a diamond; and a sub process or a more intricate series of steps that need to be followed but can be summed up in a single name, typically shown as a rounded square.

Here are these symbols in image form.

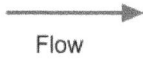

A **flow** symbol, or an arrow, shows the transition from one process to another. It is used to logically link processes together and explain the order of operations for these processes. That is to say that the arrow shows an output for a process and an input into another process and the order in which those processes take place. Some academic textbooks will discuss the fact there are differences between data flow arrows and physical flow arrows. For our purposes, this is just in indicator of saying one process flows into another. Think of this like an assembly line indicator showing which way the assembly line moves and when items should be moving between stations that do something to the item.

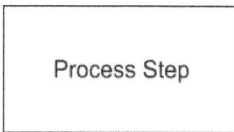

The **process step** symbol is typically shown as a box, or a rectangle. This is a great symbol because it is very easy for you to draw in person on a napkin or a whiteboard but is also easily created in any manner of electronic documents such as a word processing document or even a PowerPoint presentation. A process step has a name. There are many formal rules as to how you have to name your process step. I prefer to keep this as simple as possible. Try to name your process step with a verb – "get raw materials", "apply first coat of paint", "put order in the shipping box", "call customer back." These are great examples of obvious steps that someone could see and know immediately what to do. If you find yourself using a noun – "new product development" - then you are likely describing a process that can be broken down further into steps (and be represented via a sub process symbol as described below).

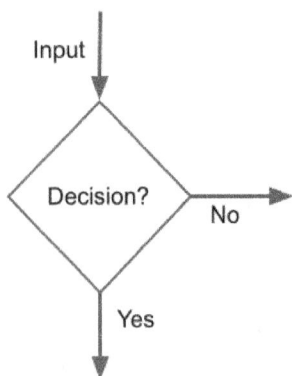

A **decision point** is a split in the path, a fork in the road. Often, a business process is not strictly linear. There are contingencies that you take. For instance, if a product does not pass a quality inspection, what do you do? Or if a customer requests a custom order, do you have a different fulfillment process? Do you ship everything in the same size shipping box to your customers or do you pick the size of the box based on the type of order your shipping? A decision point is shown as a diamond with one arrow coming in and multiple arrows coming out of the diamond based on the answer to the question you are deciding. A decision point often has a name that involves a question mark – "Customer satisfied?", "Expedited shipping?", "Customization required?". Most decision points that I end up writing resolved to yes or no questions. This binary question is incredibly important. It really simplifies the decision-making process that is required to understand the business process I'm outlining.

A **sub process** is useful when creating higher-level process diagrams. For instance, if you work in manufacturing, you likely have a process for transforming the raw materials and a separate process for inspecting those raw materials and finally a third process for packaging and shipping those raw materials as a finished good. Each of these processes needs to be understood at high-level so you can arrange your warehouse in a corresponding manner. Yet each of these are themselves highly detailed processes. As a result, when you find yourself building a process flow diagram and you need to use a noun as one of the steps other than a verb, use the sub process indicator to alert the reader that there are more steps inside of this box that you have not exposed at this time in this process flow diagram.

Using these four simple symbols, and very basic naming rules, you now have the ability to document nearly any process inside of your business. By documenting the processes that you currently follow, it becomes obvious where you can make changes and improvements. You also have a way of communicating an abstract idea, the process you take within your business, as a drawing that everyone can now read and understand. When you then choose to use technology to speed up or automate a part of your process, you now know exactly what box you are changing and how your change might affect the process.

You will also be alerted to the fact that some technologies seek to not only change how you do your process but also seeks to change a process itself. This is one of those areas where your business will feel the technology is not working for it if you inadvertently change your process through adopting a technology and do not know where in your process you changed the steps that you are doing. These basic rules to document a business process will help to ensure that you understand what is happening inside a process and to be knowledgeable about any changes you are making to the process and why you're making those changes.

Chapter 11

THE TWO CRITICAL TECHNOLOGY FRAMEWORKS

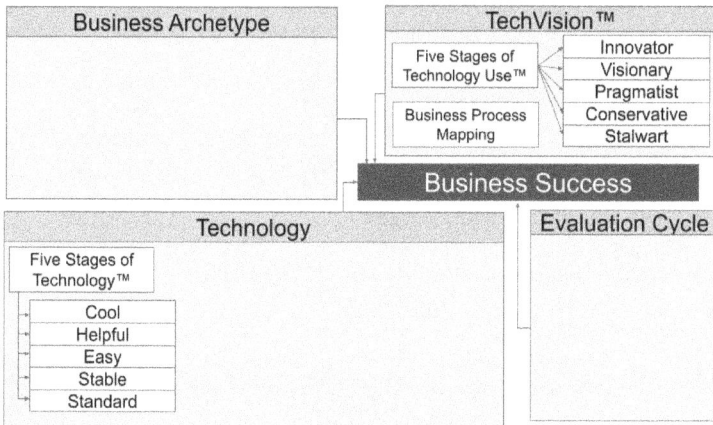

Business Archetype		TechVision™	
		Five Stages of Technology Use™	Innovator
			Visionary
			Pragmatist
		Business Process Mapping	Conservative
			Stalwart

Business Success

Technology		Evaluation Cycle
Five Stages of Technology™		
	Cool	
	Helpful	
	Easy	
	Stable	
	Standard	

In this section, we'll talk about two theories around technology and your company. There are two frameworks for looking at the fit of technology with your business. The first of these is called the Five Stages of Technology Use™. This framework looks at your business and how you as a business look at technology. Likewise, there is a second framework called the Five Stages of Technology™ that focuses on the technology itself regardless of your business feelings towards technology. Both of these frameworks need to be mastered to ensure the technology is actually going to fit your business and alignment with your TechVision™.

THE FIVE STAGES OF TECHNOLOGY USE™

The Five Stages of Technology Use™ is a framework of looking at your own business' technological prowess and understanding how you wish to use technology. This is very similar to building a culture in your business. The Five Stages of Technology Use™ is intimately human in its application. Because every individual has their own fit with the Five Stages of Technology Use™, you and your employees may score differently on this aspect of your CHESS™ Profile.

There are five areas of the Five Stages of Technology Use™. It is very common for one individual to move between the five stages depending on the area of application of technology. Do not assume that simply because somebody falls in an Innovator category that they are always going to be an Innovator for every aspect of technology inside the business. These stages are Innovator, Visionary, Pragmatist, Conservative, and Stalwart. Let's cover each of these in-depth.

Innovator

The Innovator is the first one in any new technology in a particular field. An Innovator focuses on the future potential of technology. But an Innovator also focuses on technology for the sake of technology itself. An Innovator is often the person or company that has new technology lying around, sometimes in various states of disrepair. An Innovator thrives on new opportunity and new possibilities.

But an Innovator also has a downside. Many new opportunities do not pan out. This leads to technology spending that only provides insight and learning, but no other business value. Just like a tinkerer's workshop is filled with half-finished projects and machines in various states of repair, so too an Innovator's business may have the feeling of being a bit chaotic.

Visionary

The Visionary is like an Innovator in that they prefer to get in near the ground floor of any new technology. Yet a Visionary has one key difference. A Visionary looks towards the business value first of a new technology. It may not be very obvious what the business value is in a new technology, but the Visionary is willing to invest time and money to find a strategic competitive advantage in that technology. It is important to recognize that a Visionary often has to put additional blood, sweat and tears, as well as, money into a technology to make it work. But the

Visionary reaps the reward of being unique in the industry and having a process that their competitors cannot match.

A Visionary should expect to spend more on technology than the offering price. The Visionary has to integrate the technology, often paying out-of-pocket to modify their own systems to work with the new technology. But the increased technology expense should be met with an increased sustainable profit margin as they become the first in their industry to offer this new technology to customers.

Pragmatist

A Pragmatist differs from a Visionary in that they wait for technology to be mainstream. A Pragmatist is the first of the adopters of new technology that has crossed the chasm from idea to actual business value. A Pragmatist wants something to be easy to use, work out-of-the-box as a complete solution and provide business value.

A Pragmatist may or may not be price sensitive. A Pragmatist is looking for the business value and is willing to pay the fair market rate for the complete solution. The complete solution is important here. A complete solution is one that can be installed, integrated, used, and updated within a standard process for that technology. While there may be updates and minor changes to the product, a Pragmatist looks for a solution that includes the full cost of getting started and would be unwilling to hire third-party contractors to modify their systems if they did not know this was a part of the package.

Conservative

A Conservative is the middle group of adopters in a new mainstream product. A Conservative wants to buy from the market leader. And they are looking for the complete all-in-one solution. A Conservative is often drawn to a platform technology. A Conservative wants to buy the technology and then plug in add-ons from the ecosystem that works with their base product.

The downside of being a Conservative is that the strategic competitive advantage has largely been lost. This is because multiple groups of buyers have already had a chance to use this technology and roll it out across the industry. By the time an ecosystem is built around a product with ready to use plug-and-play features, the product itself is fairly standardized. A Conservative reaps the reward of being a part of the mainstream but does not overinvest in new technology that may lead to wasted technology spending.

Stalwart

A Stalwart is the last company into a new technology trend. By the time the Stalwart adopts technology, it is commonplace in the industry. A Stalwart may feel that they are playing catch-up in certain aspects of their business simply to keep pace with the rest of the industry.

It is not always bad to be a Stalwart. In many areas of business, the Stalwart reaps the business functional benefit and modifies their business to keep pace with the industry without having to pay for the industry to learn and grow. While they may not be at the cutting edge of industry, a Stalwart often has technology that works well for their business and very minimal wasted technology spending.

THE FIVE STAGES OF TECHNOLOGY™

There are five stages that a technology goes through. Similar to the Five Stages of Technology Use™, which describes the view that a business has with regards to how they want to use technology, there is a Five Stages of Technology™ curve as well. Technology as it rolls out across a new industry goes through a number of changes as it morphs from new idea to accepted solution.

If a company does not understand where on the Five Stages of Technology™ a technology falls, it is highly likely to adopt a new technology too early and waste money on technology that does not meet the business needs at the time. Often, a media publication will suggest the new cool thing in an industry and businesses will seek to adopt that new cool thing. Yet, these businesses do not understand the full cost of eking out business benefit from what may simply be a futuristic idea today. This leads to disappointment and technology not fitting the business need.

Every technology starts out as a new idea. In this stage, the new technology is simply the plaything of technologists who marvel at the technology changes for the sake of the technology. This is not where most businesses derive value. As the technology matures, business value is discovered and systems and processes put in place to extract the business value from this technology in a repeatable manner. Once this process is solidified, the product becomes a mainstream accepted technology and everyone piles in.

However, most technology fails to cross the chasm. There is a whole book about this called *Crossing the Chasm*. I highly suggest that you add it to your reading list if you are serious about understanding technology in a business.

The chasm is the point at which technology moves from the realm of the business who is willing to put in additional work to get the business benefit to the realm where the business benefit is baked into the complete solution that can be used by anyone. This is a key point in time in a technology's lifecycle because it means that the technology has moved from an idea to a product. Moving from an idea to a product requires about a nine times increased investment over the cost of developing the idea. At this point, the increased cost is no longer borne by the business that seeks to adopt the technology but is instead baked into the price of the technology as it enters the broad market.

Let's discuss the Five Stages of Technology™. The Five Stages of Technology™ can be remembered with the acronym CHESS™. They are Cool, Helpful, Easy, Stable, and Standard.

Cool

Cool technology is at the cutting edge of what is possible. Cool technology is cool. The press will tout it as the next great thing. For instance, when the Internet was first being discovered, the press began to envision a world whereby one could publish anything online and all the information would be available. This was of course unfathomable to the average populace, most of whom didn't not even own a computer.

Cool technology is cheap. This is because you are buying only the technology. This would be akin to buying the engine for your car and just that. No tires, no seats, no interior, no body. The technology is cheap to buy but it requires an intense commitment of your time to turn it into something that is usable. For this reason, Cool technology is not easy to use. In fact, the hurdles you have to overcome to even use the technology is often a part of the allure of the technology itself because not everyone gets to be a part of the technology users group at this point in time.

Yet, because of these hurdles and the fact that you are not actually buying the end product but rather the basic

THE TWO CRITICAL TECHNOLOGY FRAMEWORKS

technology, perceived usefulness is very low. But perceived usefulness is, at this stage, highest to Innovative individuals. That is because the value of tinkering with the technology is itself worth something to an Innovator. This is the key disconnect between Cool technologies and business value of these technologies. This disconnect is a primary source of wasted technology spending for most companies. Innovators want to use Cool technology to learn how it works; your business often must pay the additional cost for the Innovator to learn the Cool technology without getting any business benefit back. This is a good way to keep an employee happy, but it is a lousy return on invested capital for the business. At this point in time, the strategic value to your company of this technology is at its lowest.

Helpful

A Helpful technology has moved from the realm of being just the technology to technology now that actually provides benefit in some way to your business. Helpful technology is expensive. This is because you now have to buy the technology and you also have to unlock the value in the technology. Often, a Helpful technology is one that must be bolted into your business by a third party or by your own technology department. As a result, there is a large investment of your time to make this technology helpful to your business. Because of the investment in time and money to turn the technology from a Cool technology into a Helpful technology for your business, the ease of use of this technology is incredibly low.

However, to the company that recognizes the value in a new technology, this technology is seen as highly useful. Because it is hard to integrate this technology into your company, it is expensive for your competitors to copy. It provides value to your company. This is a strategic competitive advantage to your company. In fact, if your company is going to make the investment of time and money to eke out the benefit of this technology for your customers, this is the point at which the strategic value to your company is the highest. You will be the first and likely only company in your industry for quite some time to have this benefit offered to your customers and you can use it as a strategic marketing advantage to draw customers away from your competitors.

Easy

Easy technology has crossed the chasm from innovative idea that provides business benefit to a product that any business can buy. Easy technology is the all-in-one solution. While it may require some assembly at your business to use, it comes with instructions that work every time. As such, Easy technology is often of medium monetary cost. The cost of the technology is increased by the cost of the installation process and the full product. Yet, because the product is a repeatable process and has instructions, you gain the benefit of everyone else who has invested their own time and capital to discover what the process of using this product repeatably and reliably is.

There is a medium time cost with easy technology product. This often is the cost to integrate the technology into your business and begin using it with your new processes.

The ease-of-use of an Easy technology and perceived usefulness are at its highest in the technology curve. This is because the technology comes with easy-to-follow installation instructions and is not very expensive. It is a product. And as a product, your business can use it and receive the expected benefit that is marketed with this product. But because everyone else can purchase this product and unlock the value that the technology offers, the only thing stopping your competitors is the price and time investment. As such, the strategic value is lower in this step than the previous one and would be classified as a medium strategic value.

Stable

Stable technology has been proven in the marketplace, installed in multiple other companies, and has clear business value. At this point, it is a known technology with known outcomes. The tool has proven what it is good for, it is clear what it is not good for, and there is an accepted manner in which to use it in your business. Because of this, the monetary cost is incredibly low. This technology is a tool and, as such, is bought for a certain outcome. It may be modified if required, but there are known modifications that can be made with this tool.

There may still be a medium time cost to this tool. And that is because you might want to integrate this technology into your business or learn how to use it in your business processes. While there may be training, the technology is not quite at the point where it has entered the social zeitgeist such that everyone, even children, understand how to use it.

The technology is easy to use at the Stable stage. All the bugs have been worked out and those that remain are known with easy workarounds that one can find through searching online. The technology has become less useful as a cool factor but is now useful because everyone else is using this technology. As such, the perceived usefulness to your business is at a medium state.

But because this technology is a known product and a known tool with obvious outcomes and real value that has been demonstrated, the strategic differentiation value to your business is very low. At this point, the technology has pervaded your industry such that most competitors use this technology successfully.

Standard

A Standard technology has successfully moved beyond the realm of being thought of as a technology. As I write this book, I am using a laptop with word processing software to put my thoughts down on what will eventually be a physical book. Going back 2,000 years, I would have been considered one of the elites for even being able to know how to write and speak in this manner. Yet today, the fact that I am writing a book raises no eyebrows amongst my readers.

A Standard technology is incredibly cheap. In fact, it may not even be recognized as a technology. Printing a book was at one point incredibly expensive. Now, books are given away for free. The time cost to use the technology is very low as well. From an early age, students are taught how to read. The technology of a book is become part of the social mores of the world.

Because using the technology is taught at an early age and is seen as standard, ease-of-use is incredibly high. Within your industry, the table stakes of knowing how to do your job include being trained in the technology for your industry that is standard. Everyone is expected to know how to use technology for your industry that is standard. As a result, the perceived usefulness of this technology is very low. The technology fades into the background and becomes part of the industry and defines the industry. An example of this is a welding torch. The ability to weld metals is an incredibly powerful technology. Yet, as any welder will tell you, they are taught how to use a welding torch as part of learning the profession and indeed, the technology defines the profession. It is therefore not exciting to a welder to know how to use the technology of a welding torch.

Because everyone else is using this technology, the strategic value is almost none. In fact, it may be that the use of this technology defines your entrance into the industry itself. Therefore, it is not only of no strategic value but is what defines the industry.

To use another example of a standard technology, let us consider glassblowing. Blown glass uses a glassblowing furnace. This specialized piece of technology is really a requirement to be a part of the blown glass industry. By buying a furnace to use for blowing glass, you enter the blown glass industry. Without this furnace, you are not a glass blower. There is no strategic value in owning a glassblowing furnace because it defines the fact that you are a member of this industry.

Chapter 12

WRAPPING UP "TECHNOLOGY DOES NOT FIT MY BUSINESS CULTURE"

A s we wrap up this section, I'd like to address some common concerns that business owners have shared with me over the years. These concerns relate specifically to the common problem that "technology does not fit my business culture." There is no particular order to these concerns. I wanted to take a bit of this book to address the concerns directly. This section might help you, as well, if you have some of these thoughts after reading the previous chapters.

My process can't be mapped and is too complex for any off-the-shelf technology.

When I was teaching a class at University, a student of mine brought up a similar point to me. They were struggling with the complex process that I'd given them for a homework assignment and declared that it could not be mapped using the tools I provided. In fact, one of the key aspects of the assignment that the student was struggling with was the fact that this process needed to be mapped on multiple levels using the sub process concept.

Often, when a business owner tells me a process is too complex to map, I challenge them to consider that they are looking at a process of multiple levels of detail simultaneously. This would be akin to a business owner describing the process of pouring a bowl of cereal as getting a box of cereal from the pantry, pouring the cereal in a bowl, birthing a cow, raising the cow, inducing the cow to produce milk, pasteurizing the milk, arranging for transport of the milk, storing the milk in a commercial freezer… You understand where I am going with this example. The different levels of abstraction in the process, if not articulated correctly, can lead to what seems to be an impossibly complex process document. The appropriate way to deal with a complex process like this is to look at the highest level possible and simplify the number of steps, usually no more than five, and then break down each individual process at the next layer of detail.

In this way, you will often find that the higher-level process that your firm is operating on does actually match off-the-shelf commercial technology. It just may be that the individual steps you take in each process needs to be modified to fit the off-the-shelf commercial technology.

My business fits in multiple stages of the Five Stages of Technology Use™. How is that?

I gave a speech to a Fortune 500 company detailing the levels of technology used inside the company. It turns out that many companies of sufficient size actually have multiple stages of technology inside of them. In fact, you can find the same phenomenon even in a one-person company.

Consider if you would be willing to use a new and unproven technology to order your lunch. In the worst-case scenario, you may get a dish that you did not want or be forced to forgo lunch altogether. Yet on the other end of the spectrum, consider being asked to use an unproven technology to perform life-saving surgery on yourself. In the worst-case scenario, you die. Using this example, it is clear that even as an individual, you have many stages of the Five Stages of Technology Use™ that you deem acceptable depending on the application.

Your business will fit into multiple stages of technology depending on which of the Five Families of Technology™ that you're looking at. Depending on your industry, you may be able to take additional risks in your marketing efforts. Or perhaps your fulfillment methods allow you to take additional risk with new technology there to gain a strategic advantage and competitive advantage over other businesses. But not every business is able to wager the same risk that the same Family of Technology because of their individual limitations.

Doesn't every company want to be an Innovator?

I'm reminded of the story about a company that rented heavy equipment to construction companies. If you've ever seen the large cranes that are used on a construction site, you likely found a company in the same industry as my friend's company. These cranes are often not owned by the construction company. Rather, they are leased for each individual job. Think about how many skyscrapers are being built in your local city at any point in time. You usually count the number on one or at most two hands. To delve into innovative, cutting-edge technology to fulfill the need of building an 80-story crane to build a skyscraper may not benefit the company or the space. In fact, you probably would feel uneasy about passing by a construction site with a new, first-of-its-kind, untested 80-story crane.

It turns out, that every company makes their own cost-benefit analysis and is entirely acceptable to not be an Innovator in every aspect of your business. For instance, you may want to innovate in your marketing efforts. Or perhaps your core value chain is an area where you want innovate to provide the best value to your customers. But it is highly unlikely that you want to become innovative and try new accounting methods that are not considered appropriate by your government or regulator.

Every company does not want to be an Innovator in each of the Five Families of Technology™. In fact, many companies take pride in their ability to not be an Innovator. Consider a luxury watch maker. It is much more cost-effective and technologically advanced to make a digital watch on a manufacturing line. Yet the digital watch sells for but a fraction of the price of a hand-crafted analog masterpiece that involves hundreds of gears moving in unison to produce a masterpiece that one can wear on their wrist. In this way, not being an Innovator in a particular realm actually allows a brand to stand out as a luxury and thus gain competitive advantage.

I don't know how to identify my company's culture.

Your company's culture ultimately depends on you as an individual to define the standards, rules, and customs that make your business the place that you want individuals to work. Your culture does not have to attract everyone. In fact, your business culture should be finely crafted to repel those who would not fit.

Your business culture will change over time. Your business culture is hard to identify but easy to describe. Ultimately, culture comes from the top of a business organization. You owe a responsibility, as the owner of your business, to define your expectations and customs to your employees.

Culture does not have to be a rigid document. It is not a series of slogans nor a pithy saying that you post at the entrance of your building. Culture is day in and day out how you and your employees react to the successes, challenges, trials, and journey of providing value to your customer.

I can't map a business process into a diagram. How can I do this?

The first step of mapping a business process into a diagram is to start anywhere in the process. I have had students have success starting at the end, envisioning the final product in the hands of the customer and working backwards to discover how the product gets there. I've had businesses have success starting at their sourcing department, looking at the raw materials coming inside their warehouse as the first step. I have had consulting engagements where the business owner started in the middle, the step in which they were determining quality control and then worked forward and backwards to discover the steps to take that product and put it in their customers hands and then likewise decompose the product into its raw material parts.

There is no wrong way to begin documenting the process. In fact, documenting a process is an iterative process. You are likely going to change your process multiple times as you recall new steps that you do, even if they are unlikely to occur in the majority of cases. Your business process diagram is not a static document. It is a living document you will continue to change as you continue to grow in your understanding of the way that your business actually works.

I'm afraid that being a Conservative or Stalwart means my company cannot use technology.

Being a Conservative or a Stalwart does not mean your company cannot use technology. When I was developing this concept of mapping a business to the Five Stages of Technology Use™, I realized that some Conservative and Stalwart businesses may feel that they were cast in a role of the "bad guy". I am sensitive to the fact that Conservative and Stalwart business owners still run a successful business. And they are able to provide for their employees and for their customers. They are not the bad guy.

In fact, many Conservative and Stalwart companies are able to price their goods substantially higher than their more technologically inclined competitors. Why is this? The answer is fairly simple.

In today's world, mass produced goods, through technology, drive down the cost of production. These are commodity goods. Everyone has the same brand from the same major retailer. To rise above this homogenous crowd, one must seek out the unique. Something that is unique is by definition not mass produced.

Being a Stalwart or a Conservative in the Core Family of Technology can actually allow your business to position itself as a luxury good. If you produce only 10 items of exceptional quality every year, and you position these items as a luxury good for the ultra-high net worth market, you can sell millions of dollars per year using just 10 pieces of inventory.

Even if you are a Conservative or a Stalwart in other areas of your business, such as the Expand Family of Technology, you can still take this position and cast it as a competitive advantage. Being a Conservative in the Expand Family means, perhaps, you only deal with new clients face-to-face. Rather than bemoan your lack of online advertising, you become the trusted advisor who always responds and meets face-to-face, providing an in-person connection in increasingly digital world. This, then, becomes your avenue to charge luxury pricing to those who desire this level of service.

Do customers want my company to be anything other than an Innovator?

What your customers want from your business is likely to be unique to your customers and your business. Not every company wants to do business with an Innovator. Nor does every person.

I'm reminded of the story from one of my high technology friends. They looked at the number of places they had put their company's data over the years as they tested out new technology to gain an edge around their chosen Family of Technology. Yet what they found was that their information was scattered across multiple systems. Every company they dealt with was an Innovator but those companies often failed. As a result, my friend began to look at more stable companies for their Help Family of Technology to reduce the chance that their company's data would be lost in another company's system when that company went bankrupt.

My software vendors don't fit my business process diagram that I made for my business process.

This is a great objection and often points to one of two problems. First, your software vendor does not understand your business process. Or second, your software vendor is actually a platform in disguise.

Do not assume that simply because a company sells a piece of software that they actually understand the process that they are selling. It is possible that the software vendor you are considering does not actually understand your business nor your business process. As a result, it does not seem that the software vendor maps to a business process because they do not map to a business process. For your business, this software vendor is not good and will result in wasted technology spend.

The second probability is that you are actually considering a platform, not a simple software vendor. Often, when you are unsure of how to map a business process in software, you are actually looking at a platform of multiple Families of Technology. In this way, it may be that this platform will help you run your business more successfully. And it will also attract new clients. And it may even help you fulfill incoming orders. The platform does not look like it fits a business process. This is because it does not. In fact, the platform is the jumping off point for the beginning of multiple business processes. If your business is not ready to take advantage of these multiple business processes from this platform, you are likely wasting technology spending.

Technology vendors make it difficult to determine if they will fit my culture. Who can I ask for help?

It is difficult to determine if the technology will fit your business culture without expert guidance. The question of if the technology will fit your business culture is difficult. In fact, there are few individuals that your business can ask for objective advice. First, asking a customer or a friend who knows technology is unlikely to give you an expert unbiased opinion. Likewise, asking the software vendor gives you a conflicted opinion. The software vendor wants to sell their services to you. Further, asking a consultant who charges per hour is likely to give you a difficult and convoluted answer. The consultant has every reason to make their advice obscure and hard to understand so that they can charge more hours to detangle the mystery that created for you.

In fact, there is really only one good person to trust when determining if a technology vendor will fit your business culture. This is a fiduciary technology advisor. A fiduciary technology advisor is someone who bills on a recurring regular basis, regardless of the number of hours they spend with you helping you to grow your business. A fiduciary puts your business first. A fiduciary technology advisor is one that puts your business benefit and outcomes ahead of their own paycheck and charges you on a fair and steady recurring basis regardless of the number of hours they work to help your business. Ultimately, you should be asking a fiduciary technology advisor for help with technology for your business.

******

Technology that fits your business culture is a joy to you and your employees. Would you like technology to be a partner and a friend? Do you want the technology at work to fade into the background? Would you like to stop struggling against technology at work every day?

If you are like most business owners, you want your technology to fit your business culture. Because technology is such a key part of your business processes, it's almost like bringing on another employee to your business. You want your employees to reflect your values and your business culture. You should expect the same out of your technology.

This brings to mind a great story of a small business that I worked with a number of years ago. This particular business was helping recent college graduates to integrate themselves into the financial system and become established adults with a proper banking account structure.

Because the small business wanted to be on the cutting edge of technology and appeal to the youth market, they knew that they could not have a culture the same as a stodgy bank. This particular company needed to have everything done in a mobile app. Yet, the off-the-shelf software for their industry did not work for them.

I remember sitting with the founder of the company as we talked with one of the largest vendors providing software to their industry. As the two of us sat on a demo call and watched the technology being used by the salesperson, we realized that the technology we were seeing was not a fit for the business owner's culture.

The technology required multiple clicks and data fields to be filled out. There was no process. It was simply a giant form that had to be completed. The business owner had built a business culture around execution and defined processes. Undefined processes led to confusion. Confusion did not fit with his high achieving culture.

It became clear that this technology would never work for this company. Even if it provided the exact output that was needed, the way in which the technology worked would always clash with the culture the business owner was trying to build. This business owner avoided an expensive mistake by not adopting the industry-leading technology as his competitors had done.

By recognizing the cultural fit between his business and the technology, this business owner was able to find technology that met with his view of the Five Stages of Technology Use™. Because he had his process envisioned and had a culture that relished being an Innovator, he was able to create technology that worked for his business and gave him a competitive advantage. This particular business went on to dominate their industry and become an incredibly popular company amongst the youth market looking to gain a foothold in their financial future.

if you were able to achieve a level of clarity about your business culture and how technology would fit or not fit your culture, would that accelerate your business further? Understanding where your business fits on the Five Stages of Technology Use™ and knowing how to map your processes to technology ensures that technology fits your business culture.

If an employee did not fit your business culture, you would know and you would take actions to protect your business culture. Likewise, you have the ability to know if technology is not going to fit or is going to fit before you purchase. Would knowing if your technology is going to work as hard as your employees be useful?

If you're like most small business owners, you want all your employees, human or technology, working together in the same culture. You can understand what your culture is through your personalized CHESS™ Assessment. Go to chessprofile.com and take the assessment today.

Chapter 13

I Do Not Know What I Do Not Know

Throughout my career as a small business owner myself, I have been struck by many times that I simply did not know what I did not know. One of those times actually came early on in my career when I was looking to build a company as a startup. As was the fashion of that time, I wanted to build a social networking mobile app.

I did what any small business owner would do. I reached out to a mobile application development company for help. I actually reached out to many companies because I knew that this might cost me a fair amount of money.

Now the reason I wanted to build a social app was I wanted to grow my business and get more sales. Now if you been paying attention to this book, you know that, right off the bat, my technology was not aligned with my business goals. I was trying to build technology in the Social Family yet what I wanted was technology in the Expand Family.

I reached out to all these mobile application developers to get a quote. I end up getting quotes back as low as $2,000 and as high as $30,000. With my personal experience now, I know that this was just the beginning of the costs to build a mobile application. Even at $30,000, I would not have an application that I was able to change and modify to reach my business goals of selling more.

But I didn't know that at the time. I signed on with the cheapest mobile application developer and expected to receive a full functioning mobile application for $2,000 that would work across all mobile phones.

What I got instead was an application that looked very little like the designs I had in my head and didn't have any of the functionality that it needed to. Simple things like the ability to change a password were missing. In fact, the ability to add any sort of content to the social network through the application wasn't available.

I didn't know the complexities involved with creating a social networking app when I first started. As a result, it cost me over

$2,000 to own what is now obsolete source code sitting in some file somewhere on a computer. I saw none of the business benefit and I paid a very real business cost.

And that lesson stuck with me.

Because I didn't know what I didn't know, I had imperiled the very existence of my startup business.

WHY IS THIS CONCEPT IMPORTANT?

Many small business owners are making the same mistake every day. It's tough to be a small business owner because you are ultimately the final deciding vote as to where the resources, people and money of your business are allocated. Yet, as a small business owner, you face the Herculean task of understanding enough about the implications of the decision you're making to make the correct decision.

But this is a very complex to do. As a small business owner, you most likely are an expert in your core industry. What you are most likely not is an expert in all facets of business. Even professional business managers, folks with MBAs, frequently employ a team of advisors and external service professionals to augment them and what they don't understand.

You have incredible knowledge running your business. Don't let anyone ever tell you differently. You are doing what most business coaches, business advisors and professionals want to do. Approach the need to gain information to make an informed decision from the point of an expert in your business. Because you are an expert in your business.

That being said, you are likely not an expert in all the tactical steps that need to be done to make your business run appropriately. For instance, I always use an accountant for my businesses. I have no desire to learn the intricacies of the US tax code nor the intricacies of all the accounting methods that need to be used to present an accurate financial picture. I can go through and tell you what a transaction is for on a bank statement in most cases, but I really have no desire to aggregate all the transactions into a quarterly or monthly financial statement. Because of that, I use an expert to do this task for me.

Likewise, as a small business owner, you most likely have an attorney. This attorney will advise you on matters of law that you do not need to learn.

But if you were to avoid ever building a circle of advisors around you, you would be forced to make a decision in the absence of understanding the effects that decision. And that is a very dangerous place to be as a business owner. For instance, if you don't know how to file taxes and instead choose to not file taxes, eventually that decision will cause massive fines that could threaten the very existence of your business.

Over the last decade working with small business owners, I have found that there are four main ways that a small business owner can deal with needing to gain enough information to make an informed decision. You will likely find yourself in one of these four mindsets throughout your small business career. I am obviously partial to using more experts than none, but all four of these approaches can be valid depending on the situation and consequences of a decision.

DO SOMETHING – THE 50/50 COIN TOSS

The discovery story of this strategy

As a small business owner, you have a bias for action. This is a business phrase that means you will do something instead of sitting around doing nothing when faced with a decision.

This reminds me of a conversation I had with a small business owner at a local farmers market. This particular small business owner was producing homemade baked goods. The aroma of her bread was outstanding. You could smell it from multiple booths away. Instantly a craving and hunger overtook me. As I approached the tent to purchase one of these loaves of bread with the intent to carry around the bread at the farmers market snacking throughout the day, I was struck by the simple set up that this baker had.

Her bread was arrayed on the table in an arc. There was one loaf of bread uncovered that was clearly where the smell was emanating from. And then she had a simple checkout counter with a mobile device and card reader.

When I went to check out, I noticed that the name of her company on the bill was her name. I asked her about this because it seemed as if I had just given my money to a friend instead of to a business. Indeed, she had simply contacted the farmers market, asked for a booth, showed up with the tent, and some bread to sell. She had never incorporated a business and was operating as a sole proprietor selling food to the public.

This story repeats time and again with small business owners. Just make a sale. Just provide value. Worry about the incorporation and legal structure later.

This is a bias to do something. You will see this in many parts of your business when you really think about the decisions you are making.

Why this strategy works

This approach to do something works really well in small business. Because not knowing does not move your business forward, it makes sense to do something to attempt to move the business forward. Often newer business owners will overuse this particular action and go around doing things without any real plan.

This approach works well because it breaks the analysis paralysis trap that many small business owners can fall into. When small business owners need more information to make a decision, they can often get stuck in this trap that they feel there is always more and more information out there to make a better decision.

One of the great axioms of Colin Powell is known as the 40-70 rule. This rule suggests that you should not make a decision before you have about 40% of the information you think you need. This is because you are likely to make an incorrect decision. Yet if you have waited until you have 70% of the information, you have likely waited too long and gathered too much information such that you now risk losing the advantage that making a decision would afford you.

This same concept is also found in *Thinking in Algorithms*, a book about structuring one's mind to be more logical. The author suggests that you should make a decision after reviewing about 30% of the data with an open mind. This allows you to establish a baseline of what your options are and then quickly make a decision with having a greater then a 50/50 chance of being correct.

What this strategy costs you

Making a decision and doing something can also cost your business dearly. Because all the conventional wisdom suggests making a decision earlier rather than later retains a competitive advantage for your company, you need to also understand the probability of that decision being right.

If you are making a decision that is a 51% probability of being right, then there is a 49% probability that you are wrong. This is obvious on its face but has important implications for how your business actually turns out.

If you were to make 100 decisions and make 49 of them wrong, the human mind begins to view that you are always making the wrong decision. This is because at about 30% or so the human mind begins to kick in and apply a general heuristic that says this "often happens". Even though statistically this is not correct, the human mind is gauging this event is having a high probability of happening. You'll think "I often make the wrong decision," even though that is numerically incorrect.

Let's say you make 100 decisions, each of these decisions is for $10,000. That is about $1 million worth of decisions. By just doing something and being wrong 49% of the time, you will spend $490,000 in wasted incorrect decisions that do not pan out that way you want them to.

While you have made more right decisions and wrong decisions, a bias for action also requires you to have a higher risk appetite that allows you to shrug off bad decisions and learn from them.

Not every small business has the resources to shrug off 49 bad decisions because they made 51 correct ones. This bias for action requires you to understand you will be wasting some resources but will be moving forward faster.

Only you can understand if your risk tolerance fits this profile or if you will be consistently awake at night worrying that you wasted money and have not gotten anything out of it simply because you moved faster.

OVER EDUCATE ON THE SUBJECT - INSTEAD OF RUNNING MY BUSINESS

The discovery story of this strategy

In an online group that I'm a part of, a small business owner lamented not being the expert in every part of their business. Logan, the small business owner, wanted to know the ins and outs of marketing, manufacturing, tax accounting, and raising capital to grow a successful business. Behind this, Logan also was an expert chemical engineer that knew everything about their product that they had created and were getting a patent on. Logan lamented how their competitors seemed to be surpassing them in sales despite the fact that some of the competitors they had talked with were completely uninformed about the intricacies of the manufacturing process.

Logan was seeking advice on how to best become a tax expert while learning how to become a marketing expert. Unsurprisingly, Logan's sales were suffering in their business. In fact, because Logan didn't believe that anyone could do as good of a job as they could, Logan was quickly becoming a one person start up destined for failure.

Yet, Logan's desire to understand everything about their business is not a bad thing. In fact, many successful small business owners suggest that Logan get a basic understanding of all these areas to avoid being oversold a solution that didn't actually solve their problems.

Yet the problem that Logan had was on the verge of costing them their business. Because Logan's desire to over educate on every component of their business was leading to declining sales, this desire for knowledge was actually hurting Logan's business.

Why this strategy works

Educating oneself on the pertinent components of their business is an incredibly valuable tactic for every small business owner. Every small business owner should have a basic understanding of every part of their business and be able to make informed decisions. Nothing is worse than not knowing anything about a subject and having no baseline for what competent performance looks like. This is a recipe for a small business owner to be swindled by an unsavory provider or taken advantage of by an employee who should know better.

Yet, too many small business owners fall into the analysis paralysis trap. If a little knowledge is good, as the thinking goes, then more knowledge must be better. If more knowledge is better, then becoming expert must be the best. This seemingly sound logic contains within it a flaw that can fundamentally derail your business.

There is a well-known effect amongst experts that many individuals who are actually at the top of their field suffer from incredible imposter syndrome. This is known as the Dunning-Kruger effect. This is because knowledge is a journey. Not knowing anything on a subject allows an individual to quickly learn and become what they feel is a master of that subject. Yet, this rudimentary knowledge is often only the beginning of understanding object or subject. As one increases their knowledge on a particular topic, they quickly understand just how much they do not know. It is this discovery process of additional knowledge that the individual did not know they did not know that leads the small business owner down the trap to decreasing sales at the same time as they seek to become an expert in every aspect of the business.

This particular idea of educating oneself to have a basic understanding works well. You as a small business owner need to have a basic understanding all the fundamentals of your business. This is a particular area where the basic education is likely what is the most beneficial to you as a small business owner.

Understanding the basics allows you converse with experts. It gives you the context to understand what the expert is saying.

And in many cases, it may allow you to identify false experts. Identifying false experts may save you from following the wrong advice, saving your money and your business.

Striking the right balance between education to have context and not becoming an expert in areas that are not your particular strengths is the challenge for the small business owner.

What this strategy costs you

As Logan found out, becoming an expert in everything in your small business could cost you the business itself. While becoming expert in every facet of your business may allow you converse at a very deep level with all your service providers and know that you are right, it fundamentally does not allow you to run and scale your business beyond yourself.

Over educating yourself on every component of your business costs you the ability to make your business operate independently from yourself. Tactically, this means that you cannot take a vacation because your business is unable to operate without you showing up every day.

Falling into the trap of thinking you have to over educate yourself and become the expert stops you from scaling your business by hiring others who are actually experts. At the root of this problem is a fear of losing control to others who are experts. Yet, trust in your employees and your service providers is a core component of your small business success.

Lastly, becoming the expert in every facet of your business stops you from running and trying your business ideas. Launching an imperfect product, one in which you are not the expert in every aspect of your business, allows you to prove product market fit faster. You avoid wasting years becoming an expert in all aspects surrounding your idea if the core idea itself will never be a successful business.

GET AN ADVISOR OR CONSULTANT

The discovery story of this strategy

My first time working with an advisor in my small business was with George, an IT expert in software development systems. I was a very good software developer early on in my career. Yet, I faced challenges as my business started to grow to thousands and thousands of people using my app every day. George was an expert at mobile application development and quickly identified multiple problems with the application that I built. It was a humbling experience to watch George rip apart a product that I had built on my own.

George came from a long background in technology. He had run very large products for Fortune 100 companies. Further, he was one of the preeminent software developers in the Los Angeles area. What I thought was going to take months to redo, George did in a weekend. And rather than just changing around my code, George fundamentally altered how our company developed our product.

Over the course of the few months the George was with us, he provided processes that major companies were using and changed the quality standards and expectations that we had of our software developers. In fact, George made it possible for us to later scale our product to be one of the biggest products in the United States in our market. George gave us the collected wisdom over his decades as a software engineer and provided it to us for relatively modest fee, considering that it would have taken us decades to learn the same lessons he had.

Why this strategy works

One of the most valuable skills that humans have developed is the ability to hear another person's experience and place ourselves in that same scenario to learn the lessons of that story without having to be present in the story itself. This is in essence what an advisor or a consultant provides to your business. One of the best ways of gaining context to make a decision in which you do not have all the information is to borrow the information and stories from somebody else who has been there before.

Fortune 500 companies look for the pattern matching that an advisor brings to your company in their most senior managers and directors. The biggest enterprises in the world place a massive monetary premium on individuals who can come into a situation, understand what is happening, and say "I've seen this movie before, I know how it ends". Understanding a situation and seeing ahead 2 to 3 years in the future allows an advisor to course correct a business owner before calamity befalls the business.

Often, individuals with this skill set are compensated at $750,000 per year or more by the large Fortune 500 companies. As a small business owner, you should expect to pay less than this, of course. However, you should also recognize that by paying less, you will receive a fraction of your advisor's time dedicated to your company. A good advisor is easily worth $1,000,000 or more per year in the total value they bring to all of their clients.

A good advisor will bring a skill set to a business that is unique to the scenario which that business is likely to encounter. In my own practice, I bring a viewpoint of technology and business strategy that allows companies to unlock the business of their dreams through technology. This is a particular area where I am the world's expert because of the work I have done both academically and in the field with small businesses. Other advisors may bring an intense knowledge of capital markets and raising money for manufacturing or expansion. Other advisors may bring a distinct knowledge of how to expand internationally and become a global company. Every advisor has their specialty that they can add to your business. You can

PERFECT TECHNOLOGY SECRETS

benefit from their experiences without having to live multiple lives to get decades of experience in each particular field.

What this strategy costs you

Hiring an advisor for your business is not cheap. Yet as the old saying goes, "if you think a professional is expensive, try an amateur." I actually experienced this firsthand when I hired an amateur because I thought a professional was too expensive. Instead of paying hundreds of dollars an hour to an individual that I knew could do the job, I gambled and paid tens of dollars an hour to an amateur. After two months of waiting for an outcome, I cut the contract with the amateur. I had wasted thousands of dollars on something that would've taken a professional an afternoon to advise me on. Yet, worse than wasting the money was the fact that I wasted months of my time that I could never get back.

In a hyper competitive business environment in which your business no doubt competes, every day matters. Waiting a quarter for a subpar outcome is not ideal. Even worse is pursuing the wrong direction for your business for years. It is time you will never get back and could imperil your business.

Using someone's experiences to accelerate your business is not free. Those individuals have invested their time and their life understanding the lessons they are now teaching you. Customized advice to your business and your situation is not cheap. Yet when you look at the income you could have in a year, the impact is large on your business. Modifying your business to avoid a negative outcome or have a greater probability of locking in a positive outcome over the next year is worth the money for a good advisor.

IGNORING IT

The discovery story of this strategy

The last way to deal with the problem in which you, as the business owner, have no knowledge and have to gain context to make a decision is to simply ignore the problem. While this sounds crazy, many small business owners simply ignore the problem in many facets of the business where they just don't have time or the knowledge to fix the problem.

I was in a butcher shop a number of years ago and walked into the restroom while waiting for my turn at the counter. While standing in the restroom, I noticed an abundance of sticky notes stuck to the various appliances in the walls in the bathroom. One said to "hold handle down all the way and count to 10". The other said, "don't use the cold water, it doesn't work". Yet another sticky note said, "more toilet paper is behind the mirror". Finally, there is a sticky note above the handle of the door that said, "just close the door, the lock doesn't work".

If you're like me, you have very large misgivings about using this particular facility now. Further, this may reflect negatively on the business owner because they can't maintain a bathroom. If a seldom used room in the butcher shop is this much in disrepair, in what quality do they keep their cutting tools. Are their knives sharp?

But think now about how you as a business owner deal with problems like this. Do you have sticky notes in your store? Do you have processes held together by duct tape or asking your employees to do something extraordinary? Perhaps you are like some small business owners who don't understand the tax code and have simply opted to not file taxes.

Ignoring a problem in your small business sounds absurd. Yet almost every small business owner at some point chooses to ignore a problem because they don't know how to fix it.

Why this strategy works

Surprisingly, ignoring a problem can work. For instance, in the butcher shop that I went to, perhaps the normal clientele didn't care about the sticky notes. Maybe there were stories behind the sticky notes that the butcher and his customers shared. Maybe in your business there are processes or sticky notes on walls that have been there for years, and nothing is falling down.

Not every problem needs to be solved in business. Sometimes, you just have to ignore the problem. By ignoring the problem, sometimes you find a solution or a new process. Or maybe you discover the problem wasn't nearly as dire as you thought it was in the first place.

What this strategy costs you

This is a valid strategy, but it shouldn't be pursued all the time. By ignoring problems instead of fixing them, your business becomes fragile. The butcher may have failed health inspection had they been inspected. This could have led to their store being closed down until they fixed the problems. Perhaps your business has problems as catastrophic as this that you are ignoring that would imperil your ability to function if a governmental authority came in and found out.

Or perhaps you have broken processes that you work around and you train employees to work around. But a new business owner would not purchase your company as a result. You may read this passage and think "But, I'm not selling the business." But you've built so much sweat equity into this business that it could be the sale that allows you to retire. If the broken processes that you didn't know enough to fix stop the eventual sale of your business, then you have been ignoring them and imperiling your retirement as a result.

Chapter 14

PLAN DO CHECK ACT – THE EVALUATION CYCLE

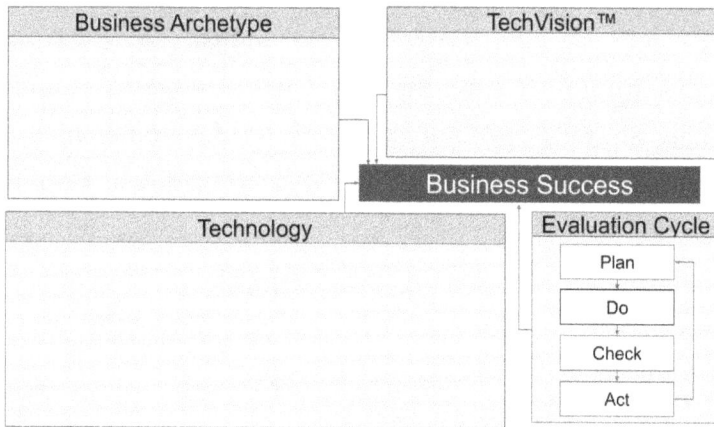

Business Archetype	TechVision™

Technology	Business Success
	Evaluation Cycle

Evaluation Cycle
Plan
Do
Check
Act

W hen I was looking for a repeatable process to describe how companies should think about adopting technology for my dissertation, I found many convoluted processes that required the business owner to study in depth how the process was run. For instance, one method required the business owner to gather experts from their industry to interview them around the ideal process for a part of their business. Now I don't know about you, but I don't often have a group of 10 or more industry experts on call for every small problem I have in my business.

I needed a process that was lightweight. "Lightweight" is a business term for a process that you can do easily, that comes naturally to you as a human being, and also doesn't require you to have advanced training in the process itself. In other words, it needed to be common sense.

Early one morning, I was reading through an operations textbook from my university days. I was reading through the quality control section and discovered the most common sense method of making a decision and then evaluating it. I realize that this repeatable process could describe how companies should think about adopting technology for their business. It would allow them to make the right decision or the wrong decision and then course correct as needed.

The process that I found was called the "Plan Do Check Act", or PDCA, process. This process actually comes from W. Edwards Deming, the father of modern quality control thinking. The original use of this process was for quality control on assembly lines and making minor tweaks to the manufacturing process. Nowadays, it is used for every manner of business decision where a process needs to be followed so the company can learn. It is especially useful for small companies that cannot have the intense overhead or extra steps associated with other more formal methods.

There are four steps in the PDCA cycle. It is a cycle that is continuous. That is to say once you get to the Act step at the end, you should begin to plan your next change and plan for your next phase of learning. There are four discrete phases in the PDCA cycle. These are Plan, Do, Check, and Act.

WHAT DO YOU DO DURING PLAN?

During the Plan phase you should be preparing an objective or end state in the processes that you want to change. When you are planning, you should evaluate all potential options available to your business. This means that a whiteboarding session usually works the best. This is because a whiteboard allows you a large space to get down all your potential ideas. During the Plan phase is important to note that there are no bad ideas. Even if someone brings up an idea that is obviously a bad one or unachievable, write it down. It will encourage discussion around why the idea is or is not possible and may reveal future improvements that you could make that are not possible for you to achieve today. During the Plan phase, you want to make note of any resources or time constraints that you may have. To make a change, ask yourself what resources do you actually have available or could you make available and how important is this change to your business.

The Plan phase will end when you have selected one particular objective that you are going to attempt to accomplish with this PDCA cycle.

WHAT DO YOU DO DURING DO?

During the Do phase, you execute on the plan that you have laid out in the previous step. This may be as simple as changing a small configuration or adding a new step to make a new manufacturing process. It could be as complex as a three-month long project. It is unlikely that your implementation will take much longer than one quarter. If it does, I suggest breaking apart your implementation into multiple cycles of the PDCA process.

It is important that you fully implement your change. In other words, you have to do what you set out to do to completion or accept that this cycle must be aborted early to learn from your mistake. Implementing a process change only partially and then becoming distracted with a new proposed change is not a recipe for success. In fact, this is what most small business owners end up doing. It is why they fail.

If you cannot do the entire change that you expect to do but continually get distracted by the next potential change you could make or the next silver bullet result in your business, then your business will likely end up amongst the failures for that year. This is because continually switching your focus without ever fully implementing a change is a recipe for failure.

Once you have fully implemented the change that you expected to do and have not gotten sidetracked by another shiny object, then you move on to the Check phase.

WHAT DO YOU DO DURING CHECK?

During the Check phase, you evaluate the expected outcome of your change against the actual outcome of your change. If you refer back to the earlier chapter about setting business goals, you know that a good objective should have been written with a measurable quantifiable result in a period of time. The Check step allows you to gauge the quantifiable metric that you were looking at. Based on the outcome of your change, the Check step will show if there was an improvement or if the change ended up being detrimental to your business. Over a certain period of time, you can then understand if this was a positive or negative change to your business and Act accordingly.

Once you Check that the change either was beneficial, neutral, or detrimental to business, then it is time to Act on your plan.

WHAT DO YOU DO DURING ACT?

During the Act phase, you document your learnings and roll out these changes to the company broadly. For instance, if you are a single person company in the business coaching sector, it is likely that you try the change with only one, or perhaps two, of your coaching clients. During the Act phase, you would then implement this change across all of your coaching clients if it was beneficial in your test.

Your new change becomes the new baseline against which you measure future objectives. If you think of the PDCA cycle as going up a hill, the Act step is where you take a breath and plan your next ascent. Once you have a new baseline established through rolling out your change to the entire company during the Act stage, you begin the cycle again starting with a new Plan.

HOW DO I APPLY PDCA TO MY BUSINESS?

The PDCA process can be applied across the entirety of your business. Every change that you make should include some aspect of this lightweight decision-making cycle. For instance, in marketing, you can do an A/B test using the PDCA process. In sales, you can test a new sales process or sales script. In manufacturing, you can run a limited-edition trial batch with your changed process. In nearly every aspect of your business, you can use the PDCA process to make incremental improvements around your existing processes.

Within technology, the PDCA process is especially important. This is because often technology must be used prior to adoption across the whole organization. When you consider technology, look for a demo period or a free trial. This is something most software manufacturers will provide. This free trial period is exactly where the PDCA cycle can help your business understand if this technology should be adopted and used across the entire organization.

Combining the idea of business process mapping that we previously covered along with setting business goals that we previously covered, we can see how the PDCA cycle allows us to trial a new technology inside of our business and understand what the result will be if we roll this out to the entire company.

While adopting technology using the PDCA cycle means that some technology will ultimately fail, this cycle in itself is actually a valid way of building your information technology paradigm inside your organization. Through the use of free trials and low license count purchases, you can limit your downside risk and financial exposure to a technology that will not help your business. You can discover technology that will positively impact your business cheaply. This means that you can be sure of the likely effect on your business if you roll out a technology broadly before you commit to the expenditure to make that purchase.

Chapter 15

Wrapping Up "I Do Not Know What I Do Not Know"

As we wrap up this section, I'd like to address some common concerns that business owners have shared with me over the years. These concerns relate specifically to the common problem that "I do not know what I do not know." There is no particular order to these concerns. I wanted to take a bit of this book to address the concerns directly. This section might help you, as well, if you have some of these thoughts after reading the previous chapters.

I don't need to gain context to make a decision

I have heard this objection and its cousin, "I don't have *the time* to gain context to make a decision", throughout the years when talking with business owners. Often, this comes from a business owner that has had prior success in one area of their business and believe that they are then an expert in every part of the business. I've often found this objection comes from a good place but the attitude behind this objection often causes the downfall of the small business owner.

I tell a story often about breaking probability. When I was younger, my friend rolled 18 sixes in a row on a six-sided dice. This is statistically unlikely to occur but not impossible. Yet the 19th roll is still just a 1/6 chance of being a six. Similarly, if you were to make 18 correct decisions in a row, you may feel confident about the 19th decision. However, you are just as likely to fail on the 19th decision as you were on the first. Gaining context is what tips the odds in your favor when making that 19th decision. That context is essential for you as a small business owner to make the right decision.

Does the Evaluation Cycle framework work over multiple years?

As I discussed earlier, while the Plan Do Check Act Evaluation Cycle is really meant for incremental changes in the short term, the same construct can be used over longer periods of time if you are looking at a discrete piece of technology. This process does work over the course of a year, which is the period of time that I suggest that you evaluate a new technology across your entire organization. This is because a one-year cycle is important for you to face the full lifecycle of technology ownership, including upgrades to the technology and your annual processes, such as year-end reports. But you must modify the Plan Do Check Act cycle slightly to make it work over a one-year cycle.

Over a year or multiple years, the PDCA cycle can be used in conjunction with key performance indicators, or KPIs. However, because multiple things are likely to change over multiple years in your company, any results during the Check phase should be weighed against other changes that have happened in the company and the macroeconomy.

Can I use different ways of gaining context depending on the problem?

In the previous chapters, I outlined four ways of gaining context to make a major decision. I highly suggest that you use multiple types of gaining context for a large-scale decision. For instance, when considering entering a new market, I would suggest that you, as a small business owner, educate yourself about the new market, speak to professionals and advisors about your options, do market research about the new market, and likely enter on a trial basis with a certain set of objectives to be met before committing fully. In this way, you use of multiple method of gaining context for a large important decision.

Now it is important to understand this is not the case for all decisions. The flipside of using too many methods to gain context is known as "analysis paralysis". It is impossible to know every component of the context you need to make the right decision. For a small decision, such as testing out a new marketing message, you can largely make the change and observe the outcome with little risk to your business. Analysis paralysis would suggest that prior to making a small tweak in a marketing message for an A/B test you would meet with your advisors, interview the market, and educate yourself on marketing to the extent of getting a professional certificate or taking a years' worth of college. This is obviously too much context and is likely to delay your decision past the point of being useful.

I have to be the expert at everything in my business. How can I give up control?

This particular concern hits very close to home. As somebody with an advanced degree and multiple certifications, I believe in being an expert in my field of business. And for a while, I too held the belief that I had to be the expert about everything my business. I needed to be the expert at marketing to craft a marketing message. I needed to be an expert in design to have a good-looking website. And I needed to become an expert in accounting order understand my finances. Yet, this was wrong because it cost me so much more than I gained by doing these tasks myself.

I subscribe to the theory of opportunity cost. In finance, this suggests that every minute can be allocated to one and only one task. The same goes with money. You can allocate a dollar to one and only one task. For every minute or every dollar you allocate towards one task, you are implicitly saying "no" to every other task that that minute or dollar could go to. If I spend two days doing my taxes, I understand at an intimate level the details of my tax filing. Yet, that is a weekend I will likely never get back. In business, these same trade-offs apply.

If you must become an expert in every aspect of your business, you are likely failing at the core process of your business. This core process is, of course, to deliver value to customers. If you take multiple days out of the week to become a marketing expert and adjust your marketing efforts, you are likely ignoring your customers and their service requests. While you may become an expert at marketing, if your business does not exist to provide marketing services to your customers, you are likely leaving your customers dissatisfied. And this has a greater impact on your business than any money you might spend on an expert to help you with marketing.

It takes too long to gain context. My industry moves too fast to wait to make a decision.

The pace of business has ever quickened over the years that I have been in the professional realm. What used to be handled on a phone call can now be done via messaging. What used to be a face-to-face meeting can now be accomplished via a virtual meeting. Gone are the days of 10 minutes in between meetings to travel from one floor to another.

Against this backdrop, you as a business owner are faced with faster requirements for every decision you make. It may seem tempting to simply give up any hope of having context to make a good decision. Instead, you achieve an endorphin high by making multiple decisions in rapid-fire. It feels good to make a decision because you get to cross something off your to do list. But without the context, you are simply flipping a coin that your decision is the right one.

A prudent business owner must look at the actions that are necessary to keep the business on course to achieve the objectives you have laid out. This means if you need to take time to gain context to make the right decision, this is more important for your business than making the decision faster. It is highly unlikely that taking another week to make the correct decision will impact your business in such a way that you cannot possibly recover from the competition making that same decision a week earlier.

I am overwhelmed at the number of decisions I need to make correctly. What do I do?

It is not easy being a small business owner. Every decision ultimately ends up on your desk, whether for the first decision or for approval of a decision somebody else is made. That "someone else" who made a decision you need to approve could be an employee, it could be an advisor, or it could be a supplier. Either way, it ends up on your desk to make the final call.

If you find yourself overwhelmed by the number of decisions you need to make correctly, you should consider which decisions you need to gain context and which decisions can be delegated or delayed. It's true the gaining context to make a decision takes an investment of time and mental capacity. If a decision must be made on something that is not important and not urgent, it is possible to simply ignore that decision. This is the appropriate use of ignoring a decision. If something is urgent but not important, that it is likely that you can make a less-informed decision because the impact is going to be very small. If something is very important but not urgent, you can use the time to gain context. And for the decisions that remain urgent and important, you as a small business owner can dive in and use the full capacity of your brain to make the right decision in the amount of time that is required.

Every coach or advisor is just trying to get my money. If they could do what they coached, they would be running their own business not coaching me.

When it comes to gaining context, some business owners will do everything possible to avoid asking a business coach for help. The thinking is that a good business coach should be in business in their chosen field instead of providing advice to the small business owner. And this is, quite frankly, fundamentally untrue.

Most business coaches I have encountered, including myself, come from a background of business success and a desire to teach. Not every individual finds joy in explaining concepts or presenting educational material to a group of individuals. Business coaches are those individuals who have been successful but also wish to mentor or advise and pass along their knowledge. Yes, there is a living to be made in business coaching and a coach who has been successful will charge a fee commiserate with their coaching and your success.

The fundamental drive behind a business coach is not dissimilar to that of an author. Having successfully climbed the mountain of business and achieved a result, the author seeks to put into words the journey that they faced and the guide to replicating their accomplishments. Likewise, the business coach does the same, but imparts their wisdom through one-on-one sessions or group coaching.

Whether it is reading a book, using a mentor, or hiring a business coach, the methods of knowledge transfer and gaining experience from those who have been on the path before you are a time tested and well honored tradition amongst every major society in every nation worldwide.

Will my customers or employees think less of my decision-making ability if we have to replace a technology that is not working?

One of my friends actually lamented when they tried a new technology for their business and on-boarded a number of their customers to it, but a few months later, the startup business shut down. My friend felt disappointed that they had on-boarded their customers to this new technology and ended up having the technology shut down under their feet. When they reached out to the customers were using this technology and had to move them off the technology into a new platform, they found that most of the customers actually were understanding. The customers expected that my friends' business was trying new things and growing and understood that not every decision was going to be successful. Because my friend had started with a small group of customers and asked them to do the minimum work possible, they were understanding that they had to do additional work to move on to the new platform.

You may get the rare customer who is upset by your business having to transition technologies. But, more likely, customers that need to transition to the new technology with your business will be understanding if not sympathetic. This is because every business goes through good and bad decisions. Not every decision is going to be correct. Just like you know this, so, too, do your customers. It is of course still important to seek to make the right decision to begin with, but do not be overly concerned that your customers will leave you if you make the wrong decision. This viewpoint simply has not been the case for my own business nor the many businesses that I have worked with.

******

Who do you turn to when you don't know? Do you want to ask an expert? Do you wish you had a guide sometimes?

If you are like most small business owners, the answer to all of these questions is yes. We can never know everything. The most successful business owners have a group of coaches, advisors, and experts that help them navigate the uncertainty of their business environment to bring about business success.

There's a great company that I worked with as a technology advisor that really exemplifies the benefits of having expert advice. This particular company is in the energy industry and is working to bring energy into emerging markets and provide the basic necessity, electricity, that so many of us take for granted to areas where the power grid is not as stable as many developed countries.

This particular business owner exemplifies the process of gaining context. Through her journey leading this energy business, and working with me as a technology advisor, she was able to turn around a company struggling to ship products and become a powerhouse in her industry.

Over the many meetings that I had with this business owner, we worked through how technology could best help her business succeed. We envisioned a future product that might change the energy industry. And then, through continued advising, this business owner was able to bring that new product to the market.

What began as an engagement where this business owner knew only the basics of technology ended with her being able to confidently interview a head of engineering for her business. She was able to gain context on technology in her business and know what she didn't know to ask the tough questions needed to interview an expert in their field and ship products on time.

Because this business owner valued her time, she did not seek to train herself on a piecemeal basis. She recognized that while formal education was an option, having a business advisor who is an expert in their field would help her retain her focus on growing her company while still tapping an expert of technology to help guide her in making decisions for our company.

Would your business benefit if you had more experts giving you guidance? Would an expert you could turn to when you are struggling to make an informed decision help you?

I have a technology advisory business that I run specifically to provide small business owners with expert level guidance and insight to help them make the best decision about technology to achieve their business goals. While I don't take every business that applies to work with me, I do look for some particular traits in the business owners that I work with.

If you are a business owner that believes technology will help your business, is passionate about serving your customers with the best products and services and is excited to team up with an expert to achieve your business success, then we will probably work well together. Go to mycto.com and fill out an application to work with me today.

Chapter 16

THE TECHNOLOGY FACTOR

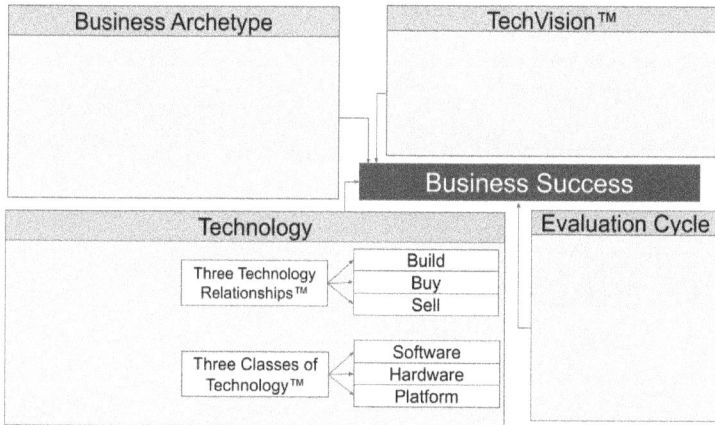

I f you look back at the beginning of this book, we outlined four variables that determine your business success. We have already covered three of these variables in depth - your Business Archetype, your TechVision™, and the Evaluation Cycle. The last variable that determines your business success is the technology itself. We have covered one part of the technology variable, the Five Stages of Technology™, earlier in a previous chapter. In this chapter, we will touch on two additional components which are the Three Technology Relationships™ and the Three Classes of Technology™. Combined with the Five Stages of Technology™, these three frameworks make up the technology factor which helps determine your business success.

The technology factor is the technology that you are considering adding to your business. The three frameworks within the technology factor help you declassify technology more easily to determine if it is going to fit with your business. We'll talk more in Section Three about actually applying the CHESS™ Framework to your business using your CHESS™ Profile and the technology you are considering.

THE THREE TECHNOLOGY RELATIONSHIPS™

There are three ways that your company is going to be working with technology. Every piece of technology will fit into one of these three relationships with your company. It is important that you as a business owner understand these three relationships and can correctly identify which of the relationships your business should use. Often, I see that misclassification of the relationship with technology ends up costing the business owner. The reason for this is that each relationship with technology carries with it a whole additional slew of context and informational overhead your business must be ready for. Let's use a quick example to demonstrate this.

Buying a car requires you to do so much more than just by a car. You are also passively accepting a service agreement with the car dealership when you have a problem with that car such as an oil change or a broken part. You end up going to the service provider or finding a certified service provider for your particular brand of car. In this way, the technology of the car itself comes with a relationship that you must maintain beyond simply using the car. Your relationship with your car's technology extends well past the day that you sign on the dotted line to purchase your car.

There are Three Technology Relationships™ - buy, build, and sell.

As I mentioned, it is incredibly important to get the relationship with technology right for your business. Simply buying a technology, instead of building that technology, frees your company from a whole host of technological concerns. Likewise, assuming that you are going to be able to sell the technology that you build signs your company up for a whole host of extraneous services that you must now provide around the technology you built initially for your internal use only.

Buying technology

The first relationship with technology and your company is buying the technology. Buying technology puts you squarely in the driver's seat and ensures the technology you are purchasing is aligned with your business. However, there are some drawbacks to buying technology which we will cover. Some of the main benefits of buying technology is that you are a consumer. You end up with a service agreement with your technology vendor. You are not responsible for the maintenance nor upkeep of the technology itself. Where it is possible to customize technology through vendor provided options, you have this capability.

However, the product you are buying is indeed just that, a product. Much the same way that going to a retailer and purchasing clothes limits choices with what the retailer has in stock, purchasing technology limits your options as to what the technology can do for you. Some of the cons of purchasing technology are the limited choice of technology. When purchasing a technology product, you are more likely going to have to change the process that your company follows around this particular area to adapt to the new technology. It is unlikely that you will be able to modify the technology sufficiently to keep your existing process intact.

Some companies find the thought of changing their process for the technology unacceptable. This may be because your company has the resources to change the technology, or it may be easier to change the technology than to change your company. It may also be that your process is one that is highly specialized and must be followed exactly. In that case, we need to look at new relationship with technology, such as building technology.

Building technology

The second relationship with technology is building technology for your company. Most companies expect to be able to buy software and then have it readily modified to fit their business processes. This is similar to many of the major software platforms available for large enterprises. For instance, a service provider like Oracle or SAP will provide not only a license and their software, but also have integration specialists to customize the software specific to the enterprise purchasing the software. Many small business owners fall into the trap of expecting to be able to build their own software without properly understanding what the technology development lifecycle, or software development lifecycle, looks like. In this way, business owners expect that they can purchase software and then modify the software through hiring a programmer from an online freelance market. This is incorrect.

Using a programmer in this manner is often an invalid method of modifying technology because it does not take into account the software development lifecycle. The software development lifecycle is a process developed by technology software specialists to ensure that the software being produced actually meets the expectations of the business. The concept of the software development lifecycle can fill a book. In fact, there are many books on software development lifecycle, academic courses teaching these concepts, and industry conferences on a yearly basis.

Underestimating the amount of work required to build software sets a company up for wasted technology spending and a product that does not fit your needs. However, there are definite benefits to building one's own software assuming you understand how to build a software or hardware development organization within your company. The software development lifecycle has a required overhead that includes understanding how to recruit and pay engineers to build your technology. While your company can unlock incredible process savings through mirroring your manual processes into the realm of technology, this is difficult to do in many cases.

One of the benefits of building your own technology is the ability to link together other disparate systems. For instance, when looking at automation tools which provide a link through APIs between pieces of software, the small business owner realizes that it is possible to link together disparate systems into a single small business IT system that could completely fulfill their needs.

For example, allowing email addresses to flow seamlessly into a customer relationship management tool, and then update your records for that customer based on the actions they take, such as purchasing an item from your website, provides a seamless automation tool to better market to your customers. This is an example of custom technology that must be built, but it can unlock powerful benefits for knowing exactly who your customer is when they get in contact with your customer service teams.

Before you begin building your own custom software, you must consider the maintenance of the software you will build. When you build your own custom software, you have now exited the publicly available market for technology. If one were to buy the chassis for a car and then build their own engine on top of it, is unlikely that a brand name dealership would know how to service your car. Likewise, when you build your own technology, there is no company or support desk to ask for help when it breaks.

Servicing technology you build means you are truly on your own. When looking at the software development lifecycle, not only do you have to make new changes to your system, but you also have to maintain the existing software. In this way, the maintenance of the system you built quickly becomes an additional expense for your company that you must plan for on a regular and recurring basis. Failure to plan to maintain your custom-built software leads the software to eventually becoming obsolete and broken. In this case, you will be hiring a new programmer to come in and look at your software who likely did not write the software in the first place. This course of action costs up to 10 times more than the initial cost to develop the software. Maintaining custom created software is not cheap.

Selling technology

The third relationship with technology is selling technology. This is a relationship that many businesses may never enter into with their technology, and that is okay. For some businesses, it makes sense to offer technology that they have built or customized internally to the broader market and external customers. This relationship with technology is often considered by large enterprises that are building technology for themselves and then offer an API into that technology for other developers to build on top of.

If you do not know what an API is, it is an easily consumable programmatic message that allows you to interact with the technology and causes processes inside of a technology company to occur. Many of the major technology companies that are in existence today offer the technology that they have built to other companies to integrate with their system.

Major technology vendors also offer software for sale in the form of product. Likewise, devices such as mobile phones are technology that are offered for sale as packaged products to consumers. It is difficult to find a "build your own phone kit" today as many consumers simply purchase a pre-made phone.

If your company seeks to get into the marketplace and sell technology that it builds, then you must fully understand what producing a product in technology entails. Building a technology product costs about nine times more than building technology for use inside your company itself. This is because there are significant additional new processes and services that are required to actually sell a product successfully in the technology marketplace.

While books have been written about properly building a technology product, I will share with you some of the high-level topics for you as a small business owner to research more if you choose to sell your technology. You must have good user experience. An easy-to-use product must be developed that looks good. Customers will not accept a product that simply works because it is the only tool available at the company. You are now competing against the marketplace of products who are all striving to be the best.

Likewise, customers now expect that there is ongoing development of a product. This means that you must now properly staff your company to continue technology development, likely forming a technology development department and hiring additional staff. This additional staff will include not only programmers but also quality assurance individuals, project management individuals, product management individuals to conduct user research, design individuals to craft the experience and style of the product and management to ensure the deadlines are met.

While there are some clear benefits to selling technology, such as taking what was previously an investment for the company as a cost and turn it into a profit center for a revenue-generating product, these are difficult to achieve. I've often seen a small business owner get excited about buying technology and then modifying it, hiring a coder to fit their business needs. Many small business owners get the additional idea that if they have a need for this modification, then this must be something that they can offer as a custom solution to the marketplace. However, you should never try to recoup the costs of your technology investment through selling the modifications you have done to the technology in the marketplace. This path of action will cause your business to derail as you are forced to support a technology product and hire additional employees to keep this customized offering in a satisfactory state in the marketplace.

THREE CLASSES OF TECHNOLOGY™

When you, as a small business owner, look at technology for your company, not only are you looking for a business need that must be fulfilled but you are also looking at the best way to fill that need with technology. In a broad study that I did of technology solutions available to small businesses, I was able to classify nearly all offerings of technology into one of three primary classes of technology. The rise of these Three Classes of Technology™ points to the various ways that you, as a business, can consume technology.

The Three Classes of Technology™ are software, hardware, and platforms. Software is of course anything written in a programming language that can be modified or changed purely in the digital space. Hardware is the physical manifestation of technology and sometimes, but not always, comes bundled with some software. Platforms are a unique blend of hardware or software that create an overlap in the Five Families of Technology™. We'll talk a little bit more about what the platform is because there are both benefits and drawbacks to your business in using platforms.

Software

Software is the first of the Three Classes of Technology™. Software technology is virtual. What this means is that it is a manifestation of a process, but not the process itself. The software that you interact with is readily and easily changed as it exists as a series of zeros and ones. Information is moved around in the virtual world through software. The divide between reality and the virtual world is evident in software. Software is able to change our views, our understanding, and our knowledge. However, it cannot compel us physically to take any further action. Software technology is often the easiest of technology to use. This is because of its immense and flexible nature.

However, software technology fails to enact actual physical changes in the world. As a result, there is what is known as the "digital physical divide". Software is only able to work in the digital world but not in the physical world.

Hardware

The second class of technology is hardware. Hardware is the physical side of technology. For instance, hardware may be a key card that you can actually touch or may be a machine on a new manufacturing floor. Hardware technology is able to modify the physical space in which we live and create actual physical goods.

In the realm of manufacturing, hardware is incredibly important because manufacturing hardware allows you to unlock efficiencies in the production line that would not be available simply through better software.

Hardware, however, is also the most difficult aspect of technology to buy, build and sell. To purchase physical hardware means that you are now required to maintain that physical hardware, often through a vendor relationship or service contract. If a relationship is not established, you may be required to maintain the hardware yourself.

Building hardware is incredibly difficult for most small businesses to do. This is because they lack the requisite physical operations knowledge that building hardware requires. Often, knowledge of the hard sciences, such as mechanical engineering, electrical engineering, and physics is required to create hardware. The hardware realm is indeed the realm of physical engineers, not weekend hobby programmers.

Some of the benefits of adopting hardware in your technology solutions is that it changes the real world of your business. By changing the real world of your business, you are able to modify the environment in which your customers experience your business. For instance, manufacturing firms may often seek to use hardware technology solutions instead of software technology solutions because they're producing something physical.

Platforms

The third class of technology is a platform. Whereas hardware and software solutions are typically focused on one of the Five Families of Technology™, a platform solution crosses either the hardware and software realms or crosses into two or more of the Five Families of Technology™. A platform solution requires you to adopt not just one piece of software or one piece of hardware from one Family of Technology but locks you into an ecosystem built around this technology. A platform can be a combination of hardware and software.

For instance, access control systems into a building that is a key card entry require you to adopt a platform. You have to purchase these physical cards from the vendor. You, then, further have to purchase the software from the vendor that allows these cards to interact with the door scanner hardware to grant access to the building. It is difficult but not impossible to simply purchase the cards and then build your own software for access. Because of this, the vendor relationship is strengthened because you are forced to use the vendor as a supplier for both hardware and software, thus creating a platform.

A platform can also exist in the realm of hardware or software alone. For instance, a platform in the software realm that crosses through more than one family of the Five Families of Technology™ is a common offering to small businesses. Consider traditional accounting software would simply allow you to create your books and ledgers for the month. That is a single piece of software that works within the Help family of the Five Families of Technology™. Yet, if that piece of software is extended to offer you the ability to invoice your customers and creates customer lists for marketing, then that software has moved into the Expand family of the Five Families of Technology™. Further, this software could then re-target your customers for online messaging through advertisements. If that software also creates invoices for payment from customers, then it is possible that software has now crossed into three of the Five Families of Technology™ by becoming Core technology. In this way, that one piece of software has now become a platform for your business.

Another similar example of this is a large technology company that offers multiple services to you. For instance, adopting an online word processing software and then using that same platform for the advertising capabilities and their search capabilities creates a platform relationship. Listing your business on their map engine and using them for your email and customer correspondence ties you into this single platform and more. One can see how adopting the services provided by a large technology company can greatly benefit the small business owner because the platform becomes a one stop shop for many of the needs of the small business. Yet, violation of this platform's rules may cause your business to lose access to nearly every technology component that it needs to function.

Platforms are incredibly helpful. The one stop shop for many of your small business needs is enticing. Yet, platforms are incredibly difficult to leave. The debate around the power of the large technology companies stems from the fact that if they remove access for a business from their platform, then that business may no longer function. The risk of being removed from a platform is an existential risk to your company.

Platforms are hard to leave, but even worse when you are forced to leave. Platforms are hard to leave because of the proprietary data format that the platform uses. Even if you were to obtain the data from a large platform, you may be forced to do significant data processing on that data to render it into a usable format for your company. A key example of this is switching email providers. While it may be possible to move email between platforms, this is not a trivial task that every small business owner is able to do on their own.

A platform is truly a double edge sword for your business. If you seek to solve your technology needs through a platform, you will have an easier time if you adopt more of the platform. This one-stop shop for your technology solutions in a particular area of your business is a great accelerant for your company. Yet, on the flipside, it is incredibly difficult to leave a platform and you are surrendering some of the control of your business to the existential threat of being cut off from this platform. Many companies can tell a story of being cut off from key technology for arbitrary reasons which imperiled their ability to reach their business goals.

SECTION THREE

Chapter 17

THE CHESS™
ASSESSMENT ONLINE

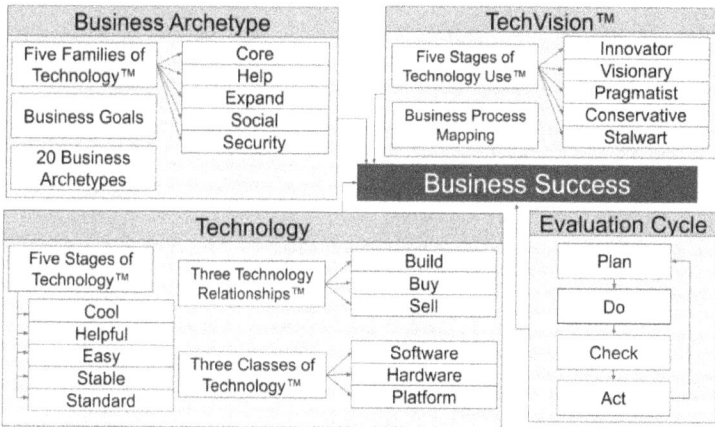

Business Archetype		TechVision™	
Five Families of Technology™	Core	Five Stages of Technology Use™	Innovator
	Help		Visionary
	Expand		Pragmatist
Business Goals	Social	Business Process Mapping	Conservative
	Security		Stalwart
20 Business Archetypes		**Business Success**	

Technology			Evaluation Cycle
Five Stages of Technology™	Three Technology Relationships™	Build	Plan
		Buy	
Cool		Sell	Do
Helpful			
Easy	Three Classes of Technology™	Software	Check
Stable		Hardware	
Standard		Platform	Act

So far in this book we discussed a systematic framework which is comprised of four key variables. One of the most valuable things about CHESS™ is that many of the components of CHESS™ can be mathematically evaluated and deterministically proven using an assessment. Because there are many moving parts and multiple conflicting business goals you as a business owner may have in your head, I knew I had to make this easier for you. I spent the last few years developing an online assessment that you can use to help you identify your key families of the Five Families of Technology™ and your view of technology which will allow you to achieve your business goals. The CHESS™ Assessment is a tool to create your CHESS™ Profile. Your CHESS™ Profile is the top two families of the Five Families of Technology™, which determines your Business Archetype, and includes your TechVision™, how your business views technology.

If you have not yet taken your CHESS™ Assessment as a part of reading through this book, I encourage you to go to chessprofile.com now and take the assessment.

I'm going to take the rest of this chapter and nerd out a little bit on some of the academic aspects of CHESS™. There are three components of the CHESS™ Assessment that point to how it can be beneficial for your company. The first of these is the general applicability of the CHESS™ system to businesses such as yours. The next is the validity of the assessment itself. And the last is the reliability of the assessment.

The applicability of the assessment to your business is really looking at how the results are going to help your business. In the case of CHESS™, the resulting CHESS™ Profile will tell you about your Business Archetype and where you, as a business owner, fit on the Five Stages of Technology Use™. Both of these data points are immediately applicable to the direction in which you are steering your business. They reveal if you are aligned inside your business, such that you will actually succeed in achieving the business goals that you have laid out for yourself. The applicability of the CHESS™ Assessment is both immediate and long-term. Because the Business Archetype underpins everything your business does, not just technology, the applicability of the CHESS™ Profile is immediate. Once you know your Business Archetype, you will be able to better understand how you approach not only technology but also joint ventures, new market opportunities, and the myriad of other offers that businesses receive on a daily basis.

The TechVision™ component of the assessment will provide you a view as to how your business looks at technology. Not only should you look at the Five Stages of Technology Use™ for yourself, but you should also include any key decision-makers inside and outside the company. These key decision-makers should also take the CHESS™ Assessment so that you understand their position on the Five Stages of Technology Use™ as well. By aligning your entire business around the same Business Archetype and the same TechVision™, you increase the probability of achieving the business of your dreams.

The next thing to consider in an online assessment is the validity of the assessment. I use the word assessment and not quiz because they are two very different things. The validity of a questionnaire that you see online should always be a major concern in applying the results to your business. In the case of

CHESS™, the assessment I have created that you see online fits an academically acceptable definition of validity. The results can and should be applied to your business.

When you see a quiz online, it is easy for a salesperson to simply set up a quiz that directs you to one of their products based on your responses. However, this is not an academic assessment, nor is it valid. Within the academic community, there is a definition of "valid" that an instrument must meet in order for its results to be considered true. The CHESS™ Assessment was designed to meet this definition of "valid".

Now, I know I've already lost some of the small business owners reading this book. You don't necessarily care about the academic definition of things, but I wanted to take a few paragraphs to include this incredibly important information around the CHESS™ Assessment. If you will indulge me one more quick paragraph diving deep into academia, we can return to the small business focused benefits of your CHESS™ Profile.

For those who are interested in the research used to create the CHESS™ Assessment, here is a brief overview of the paper I wrote. The CHESS™ Assessment is based on a confirmatory factor analysis study using factors I identified through a prior qualitative study that I ran. Factor loading was measured, and the highest loadings were then used in a pairwise comparison quantitative instrument. This pairwise comparison of questions instrument was academically tested using a chi-square goodness of fit test on two hypotheses. The chi-square goodness of fit test showed the pairwise questions using this assessment produced a result that was statistically different from the control group. The p-values for the factor loadings as well as for the chi-square goodness of fit test were all statistically significant.

What does the CHESS™ Profile mean to you as business owner? When you take the CHESS™ Assessment, the results that you get back have been specifically proven to accurately reflect where your company should invest in technology based on the business goals you have selected. The CHESS™ Profile is an accurate view of your company.

The last component of the CHESS™ Profile you should consider is the reliability. Now again, this is an academic term. Reliability means that when you take a test multiple times it continues to give an accurate answer every time you take it. If you ever taken an online quiz multiple times to get the result you desired, that quiz proved to not be reliable. In the case of the CHESS™ Assessment, the results are reliable because they will show the same view of your company given the same goals. However, in many cases, the CHESS™ Assessment will produce a different CHESS™ Profile for your business every 12 months. This is a feature of the assessment.

When you think about what is important in the next 12 months to your business, those business goals should be things you hope to achieve in the next 12 months. Once you have achieved those goals, you either seek new goals that are different, or you seek more of the same goal along certain dimensions (more sales, more customers, more employees, etc.). CHESS™ accounts for the fact that you could change business goals or could seek more of the same business goal. Changing business goals may be reflected in your Business Archetype changing on your CHESS™ Profile. Likewise, wanting more of the same business goals may result in your TechVision™ changing to reflect a desire for more streamlined processes.

Because CHESS™ is reliable across every part of your business and reflects the macroeconomic environment at the time you took it, it is expected that your Business Archetype may vary over time as you take the CHESS™ Assessment. There is an arc of Business Archetypes that closely follows the lifespan of your business. Your CHESS™ Profile will change to reflect the reality of your business in its current environment. As such, your CHESS™ Profile can be used to guide your business to achieve the goals you need to have success.

As we wrap up this chapter, I want to leave you with a parting thought about how CHESS™ can help you uncover problems inside your business. As you know, your business is more than just you as an individual. You may have advisors, or close friends and confidants, that help you to build and operate your business. Each of these individuals brings with them their own view of business success and their relationship with technology.

I have found that it is incredibly important for a business to understand the TechVision™ of their advisors and employees. Likewise, ensuring that your employees believe in and are working towards the same business goals that you are provides additional clarity to you as a business owner. If your employees are not aligned with your business, their CHESS™ Profile will reflect that. Because of the incredible usefulness of the CHESS™ system in determining where there are disconnects between you and your view of your dream business and everyone else and *their* view of *your* dream business, I highly recommend that you give a CHESS™ Assessment to each of your key employees and advisors.

I have seen business owners gain insight that their key managers are not aligned on what they think is important for the business. I've also seen business owners gain insight that the Business Archetype for each of the key functional areas of their business is different. In some cases, this accurately reflects what the goal for each functional area should be as determined by the business owner. Yet, in other cases, the business owner realizes that their company is not pulling together in the same direction because of the CHESS™ Assessment. If you don't know the CHESS™ Profile of your key employees yet, have them take the CHESS™ Assessment soon.

Chapter 18

APPLYING CHESS™ TO YOUR BUSINESS'S TECHNOLOGY CHOICES

This entire book has been working towards the goal of building a framework for the evaluation of a specific piece of technology inside your business. This is the process by which you can consider adopting and using that technology successfully. There are a number of pieces that fit together to define your probability of business success that can be achieved through using business and technology together.

In the nearby figure, you can see the four components that feed into your business success. This is the Business Archetype of your business, the TechVision™ for your business, the

Evaluation Cycle for your business, and the technology itself. As covered in this book, many factors make up these four main components inside of this formula.

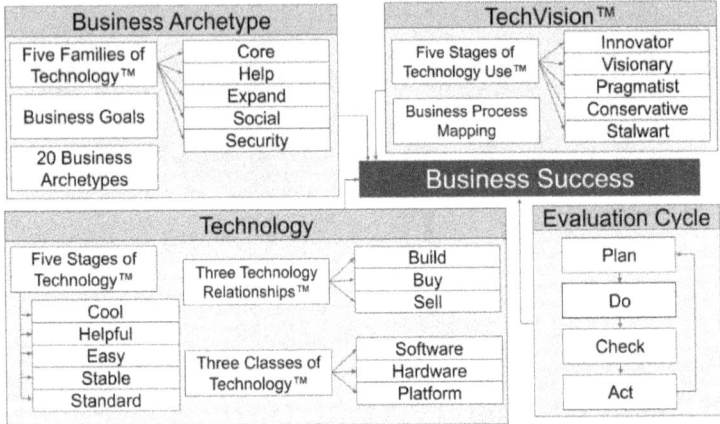

Business Archetype		TechVision™	
Five Families of Technology™	Core	Five Stages of Technology Use™	Innovator
	Help		Visionary
	Expand		Pragmatist
Business Goals	Social	Business Process Mapping	Conservative
	Security		Stalwart
20 Business Archetypes			

Business Success

Technology			Evaluation Cycle
Five Stages of Technology™	Three Technology Relationships™	Build	Plan
		Buy	
Cool		Sell	Do
Helpful			
Easy	Three Classes of Technology™	Software	Check
Stable		Hardware	
Standard		Platform	Act

Now, let's start applying these lessons inside this book to the hypothetical evaluation of a new technology for your business. I will take you through the step-by-step process that I use when looking at new technology for a client.

STEP BY STEP PROCESS

Step One

Before we begin to look at adopting new technology for a business, we must go back to the first principles of the business. Using the business process mapping that you learned in this book, map out the business process you are seeking to replace with technology. Start at a high level and build the simplest process diagram you can. And think critically around each of the boxes that you have written down. Are there sub steps inside of these boxes? If so, document the sub processes as well. A properly documented process is likely going to take you a few hours to consider all the permutations of that process. This is not to say that you should spend hours documenting every process in your company. Rather, expect to set aside sufficient time to think critically around the process as it exists today.

Step Two

Once you've documented the existing process, think critically about where the inefficiencies in this process are and what could be made better. Do not limit yourself at this point to finding a technology solution or even researching too much as to what is available in the market. The objective here is to find the ideal end state of this business process for your company.

The changes you make to the business process could be the order in which steps are done. Or could be simply making notes next to boxes saying, "this needs to be better" or "we need to improve quality". Using the lessons we learned around setting actionable business goals, modify your process to reflect the changes you want.

Step Three

Set a clear and actionable business goal and write it down. Now that you are changing a process, you are likely changing the process to accomplish the goal. This is the step in which you will clarify the business goal you are seeking to change and the desired outcome of the changes you are making. Using Chapter 6 from this book on writing clear goals, write down a clear business goal.

Step Four

Update your Business Archetype using the CHESS™ Assessment. If you have not recently taken a CHESS™ Assessment in the preceding 12 months, now is the time to ensure that your CHESS™ Profile is up to date. This is an incredibly important component of the process of adopting new technology.

The CHESS™ Profile for your business will reveal the few key aspects that technology will help with. For instance, your Business Archetype will reveal the two most important families of technology of the Five Families of Technology™ that your business should be focusing on for the next year. Likewise, your CHESS™ Profile will reveal the TechVision™ of your company which includes where you are on the Five Stages of Technology Use™.

If you have employees or other key decision-makers that are going to be implementing or using the technology you are considering adopting, ensure that they also have taken a CHESS™ Assessment and generated a CHESS™ Profile. While you as the business owner are ultimately responsible for the direction of your company and implementing the Business Archetype that your CHESS™ Profile reveals, the CHESS™ Profile of your team and employees will show where there may be disconnects between you and your employees.

This step is so vital to ensuring that technology you are considering adopting for your business fits your business goals and fits your business culture. Fitting your business goals means aligning with your Business Archetype, which the CHESS™ Profile will reveal. Likewise, fitting your business culture is aligning with your TechVision™ which the CHESS™ Profile will show for you and the CHESS™ Profiles of your team will show for your team.

Step Five

Using the Three Technology Relationships™, consider whether your business would buy, build, or sell the proposed technology. You will need to bring about the business process change with this technology to achieve your business goal. We've covered the build, buy, sell paradigm in Chapter 16 on this book. Consider which of these three areas you are likely to pursue for any hypothetical technology that you would use to achieve your particular business goal. Do not underestimate the difficulty of building your own software or hardware.

Consider that there are many aspects of the new technology that need to fit to achieve your business success. Write down what these aspects are but also pay attention to whether or not these are "must have" aspects or "nice to have" aspects. You will never find a technology that can be purchased commercially that fits 100% of your needs. Expect to have to make trade-offs between your ideal technology and that which is commercially available today.

When evaluating the difference between building or buying technology, do not underestimate the amount of time you need to put in to appropriately evaluate buying technology. The time you spend towards evaluating whether a purchase of technology is going to fit your business is time well invested.

Likewise, do not underestimate how difficult building technology is. While the initial cost may seem lower and the features may identically match what you are envisioning in your head, the total cost of ownership for building technology can be substantially higher, in the order of magnitudes (10x) higher, than buying technology.

Step Six

Evaluate the Three Classes of Technology™ and decide on your ideal class of technology. As you covered in Chapter 16 of this book, the Three Classes of Technology™ are hardware, software, and platforms. Consider which type of technology is most likely going to help your business accomplish the business goal and business process change you have laid out. It may be obvious depending on your type of business. It may also be a hybrid type of technology that includes software and hardware.

226

Even today, more hardware comes with software options that allow you to enable or disable components of the hardware and thus control your costs depending on the features that you need.

Be flexible with the ideal class of technology because it may or may not exist. We have not yet gone into the market to determine if the ideal technology we are envisioning, up to this point, even does exist in a commercially available manner.

Step Seven

Now we will take the outcome of the preceding steps and go in search of our ideal technology. If you are like most business owners, you likely began this process with the technology in mind and perhaps even a vendor that provides this technology to your company. However, now is the time to go investigate the entire marketplace. A mentor once told me that any major purchase should have at least three data points, or quotes or vendors, before a business makes a decision.

At this point, search the marketplace through the vendor connections you have, your fiduciary technology advisor, and your network. A fiduciary technology advisor at this point is especially helpful because they are up to date on the latest trends and technologies available in your market as well as adjacent markets. A fiduciary technology advisor can help you evaluate the process you are using to search for technology and ensures you achieve a significantly better outcome than simply performing an online search.

Step Eight

Once we have found a group of technologies to consider, we need to evaluate the technologies we have found. In step eight, we will employ the Five Stages of Technology™ to evaluate the maturity of each of the technologies we have found. By classifying the technologies we found along the CHESS™ scale, we will be able to understand the maturity of each technology solution and vendor we have found. Additionally, we need to map out the Three Classes of Technology™ for each of the technology solutions we have found. Finally, we need to assign one of the Three Technology Relationships™ to each of the technologies we have found to understand the properties of the technology solutions we are considering.

Step Nine

Once we have identified the Five Stages of Technology™ maturity for each of the technologies we have found and placed them in the appropriate category of the Three Technology Relationships™ and Three Classes of Technology™, we need to identify the likely winner to implement inside our business.

The likely winner may not fit all of the requirements we have had up to this point, but it allows us to make an educated trade-off between what is available today versus the need that we have to solve to achieve our business goal.

Once we have considered the trade-offs between the technologies we are considering and the business goal we are attempting to achieve to unlock our business success, we must make action and implement a technology.

Step Ten

This is the step in which you adopt, or implement, the new technology inside of your business. Vendors are signed on, contracts are agreed to, and money changes hands. Implementation begins and ends in this step.

Yet, Step Ten is the beginning of our Evaluation Cycle to ensure the preceding nine steps were indeed correct. In Chapter 14 of this book, we covered the Plan Do Check Act cycle. This cycle now kicks off in earnest in Step Ten. Through your evaluation of the technology, you will understand where, and if, changes need to be made to your business process to achieve your business goal.

It is in this final step, as well, that a fiduciary technology advisor can be the difference between your business succeeding and achieving your business goals or failing. A fiduciary technology advisor will provide you with the guidance and expert insight you need to understand when and how things can be improved. In a perfect world, there is no need to course correct nor find solutions to problems. Yet, when something does go wrong, a fiduciary technology advisor with the expertise to help your business course correct is invaluable.

These ten steps outline the process that I've used time and again with my own clients to adopt technology that furthers their business goals and unlocks their business success. I strongly suggest using this process inside your own business to ensure that you are adopting technology that will work within your business.

In my own practice as a fiduciary technology advisor, I've also refined this process and include a few steps that I've not outlined in these 10 steps above. The entire process is available in a simple one-page sheet that my clients fill out before they spend any money on technology. This very simple process on one page of paper allows them to catalog their decisions and revisit their decisions later on if they want to make a change. There were many times a client has considered a new technology and then decided against it. Later, they come back and revisited their decision, using this one-page sheet I have developed as a key decision point to determine if they should change course.

Chapter 19

RELATIONSHIPS BETWEEN THE FOUR VARIABLES

When looking at new technology, I have found that there is a relationship between the TechVision™ of the company and the technology. There is a very real reason that businesses spend money on technology that does not work. This reason is the fear of missing out on the next big thing. Trade publications in industry journals hype technology in the Cool phase as the next big thing. Companies that are struggling in the current marketplace seek a competitive advantage and begin the process of using this technology in their own business. Yet, these companies fail to appreciate how much money they must spend to take a Cool technology and turn it into a technology that provides business value and a competitive advantage that is sustainable.

COMPARING TECHNOLOGY AND PEOPLE

In the chart below you can see the interplay between the Five Stages of Technology Use™ and the Five Stages of Technology™. In each of the boxes is the reaction that a company may have to a technology in each of the categories for technology. For instance, to a Visionary company, a Cool technology has no business use. Likewise, an Innovator would look at a Stable technology and immediately dismiss it because everyone else has that technology.

When you hear these sentiments from industry publications, your team, or even yourself, this may be an indication as to where you fall on the Five Stages of Technology Use™ and where the technology falls on the Five Stages of Technology™. For each type of company, there is the maturity of the technology that fits the best along the diagonal and italicized.

But an additional important point is that the ideal technology maturity will change depending on the application of that technology inside the business. For instance, if your company has no money to spare, you are unlikely to adopt a Cool technology to attempt new marketing. However, if you have a research and development department, they are highly likely to adopt Cool technology because your research and development team is likely made up of Innovator types. Likewise, the risk of a negative outcome can affect what type of technology you are considering. Healthcare companies may likely skew towards Stable technologies as a whole, but they may embrace Cool or even Helpful technologies for marketing and sales efforts. This is because the risk of negative outcomes from healthcare equipment failing is incredibly high. Yet, the negative outcome of not making a sale of that same healthcare equipment could impact the company, but only minorly from a financial perspective. Therefore, the risk is worth the reward of having a competitive advantage through a new technology.

	Innovator	Visionary	Pragmatist	Conservative	Skeptic
Cool	Great technology!	No business use.	Too many bugs.	Too complex.	Will never work.
Helpful	Was cool, now "corporate"	Could change business!	Almost real business value.	Too many pieces to put together.	Won't deliver promise.
Easy	Helps the business.	Strategic value is passed.	Real business benefit!	Who is the real leader in market?	Technology is changing…
Stable	Everyone else has it.	Not flexible enough.	Lots of features we don't use	All-in-one solution!	We should probably use this…
Standard	Old tech.	Can't customize it enough.	It's the standard	Useful to our business	Everyone else uses it, we need to, too.

FIVE KEY ATTRIBUTES OF TECHNOLOGY

I've also broken down some key attributes for technology as it moves through the Five Stages of Technology™. These are the key attributes that I have seen many small business owners asking about. These five attributes are the monetary cost, the time cost, ease of use, perceived usefulness and strategic value.

Money and Time Cost

The money and time cost come up in conversations with business owners time and again. The first question is always "how much is this going to cost me?" in terms of dollars. This is a fairly easy question to answer when there is a product price. Yet, the total cost of ownership of the technology also includes the installation, modifications, upkeep, and decommissioning. As a result, if the technology is not yet a product, it likely has to be integrated and modified to fit each individual company's use. This requires integrators and third parties which cost money. Likewise, the more complex the installation, the more time it costs the company. The time cost of a new technology includes not only the time spent installing and setting up the technology but also includes the training and ongoing maintenance required to make the most use of that technology. Where there is significant training due to the complexity of the technology, the time cost is inevitably higher.

Ease of Use and Perceived Usefulness

Ease of use and perceived usefulness are two sides of the same coin for every technology. Ease-of-use can be best defined as the user experience. If a technology is well designed, it will be easy to use in conjunction with its intended business processes. Ease of use is most reflected in the like or dislike of technology. If the technology does not fit your business culture, it is likely that the ease of use is a factor in this. Perceived usefulness is what the technology does for its users. If ease of use is how the technology works, perceived usefulness is what it does. Perceived usefulness, it is important to note, is the perception of the usefulness of the technology that actually drives whether or not the technology will be useful. If one believes the technology will not be useful, then they are unlikely to invest the time required to achieve the benefits of using a technology.

Strategic Value

Strategic value is the fifth and final attribute for every type of technology. The strategic value is the benefit that your business gains relative to its competitors in your industry. Strategic value can be thought of as building a competitive moat through technology that is difficult for your competitors to cross. Strategic value is what you, in the business, are to be most concerned about if you are attempting to use technology as a competitive factor against other businesses in your industry. High strategic value for technology allows you to provide a marketable difference that could be the deciding factor in a client signing with your business instead of a competitor.

As you can see in the table, the five factors vary depending on the level of technology maturity. For instance, strategic value is at its highest once technology has been proven to have a business value and before the technology is either easily purchased or easily integrated into a competitor's business. Likewise, the time cost of technology is the highest at the beginning of its lifecycle and drops substantially as the technology becomes mainstream.

	Money Cost	Time Cost	Ease of Use	Perceived Usefulness	Strategic Value
Cool	Low	High	Low	Low *High to innovators*	Low
Helpful	High	High	Low	High	High
Easy	Medium	Medium	High	High	Medium
Stable	Low	Medium	High	Medium	Low
Standard	Low	Low	High	Low	None

INTERSECTION OF TECHNOLOGY, BUSINESS, AND YOU

In the following figure, I outline the intersection of technology, business, and you as the business owner. The boxes showcase the reaction that you, as a business owner, are likely to have depending on whether or not the technology fits with you or your business. If the technology and business fit together but do not fit with your personal views, or your TechVision™, then you are likely to think that the technology doesn't fit. If you believe that the business process fits your business, but you cannot find technology to do this process, you are likely to consider this as a manual process. If you like a technology and it fits for you, but you cannot find a business value for that technology, then you are likely to think that technology is an interesting toy or perhaps a fad that may come and go. It is when the technology fits the business, and the technology fits you, and you perceive the business value that you are likely to

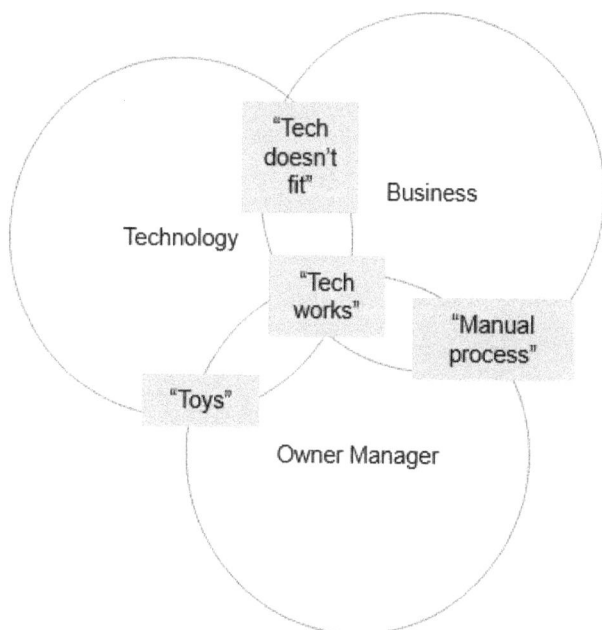

believe the technology works and likely works incredibly well.

If you find yourself muttering and thinking these phrases as you look around your business at business processes and technology, come back to this image and consider where the misalignment may be between the technology, your business, and your own personal views.

CONCLUSION

Thank you for taking the time to read this book. I hope that you have learned many new things that you can apply to your business starting today to achieve the business of your dreams. If you have not yet gotten your CHESS™ Profile, go to chessprofile.com and take the CHESS™ Assessment now.

The nagging feeling that you had when you pick up this book that your business could be better should now be resolved. You should now know how to diagnose what might be going wrong in your business. You now have a framework called CHESS™. Your framework includes four key variables that you can look at to determine how to build the business of your dreams with perfect technology.

These four variables are your Business Archetype, your TechVision™, the Evaluation Cycle, and technology you are considering.

You now understand how to measure your view of technology, where to invest in technology inside your business and how to evaluate technology to see if it will be a fit for your business.

Perfect technology makes it possible to unlock the business of your dreams. You now have a framework and the tools you need to build the business of your dreams.

*****

As you were reading this book, did you find a new way of looking at your business? Did you identify a problem and think "I should fix that?" Did you find some insight that would help you if you just applied to your business? Did you identify that there is work you could do in your business to use the perfect technology to unlock the business of your dreams?

I have one question for you if you answered yes to any of the questions above.

Would you like help with that?

I run a fiduciary technology advising business focused on small businesses like yours. I help business owners implement the perfect technology across their business to unlock the business of their dreams.

Not every business is a fit for my services. As a fiduciary technology advisor, I become involved in your business. I become a resource on your advisory board as you grow your company. I can be intimately involved on a daily basis or simply check in once a quarter. Depending on the needs of your business, my advisory position to your company could be very involved or on an as-needed basis.

If you read this book and you thought, "Wow I'd like to work with this advisor", then I think it's worth us having a conversation about what working together would look like.

Because technology advisory and being on your businesses advisory board does require a commitment on my behalf to your company, I cannot take every business that applies to work with me. Yet, I have a personal goal to help every business owner that applies to a part of my program as much as I can, even if they are not a good fit to work with me personally.

I want to help your business. I really enjoy working with small business owners and helping them realize the best technology for their business. What a dream business looks like to you may be different from the next small business owner that I work with. But I craft my advice to your business specifically, tailoring my knowledge to your situation.

When we begin an advisory relationship and I start to work with your company, I bring the full suite of my experience and the tools I have developed over the years to help businesses make the best use of technology. We will do a CHESS™ Profile for your business. I will then walk through your profile with you and we will derive a perfect technology strategy from it. Using your CHESS™ Profile and the CHESS™ Framework in conjunction with your perfect technology strategy, your business will adopt the best technology to achieve your business success.

Do you already have an idea of what business success is in your mind? Would you like me to help you achieve that?

If you answered "yes", go to http://www.mycto.com and get in touch. I look forward to hearing from you and helping your business.

www.ingramcontent.com/pod-product-compliance
Lightning Source LLC
Chambersburg PA
CBHW020159200326
41521CB00005BA/192